Living off The Grid

A Guide on How to Live Off the Land and Become Self-Sufficient Through Homesteading

Contents

Introduction

What is living off the grid? Take a minute to think about your definition. Write down the things you've heard about living off the grid and keep it handy but set it aside for a moment because you will get back to your description in a few minutes.

You bought this book because it is up to date, easy to understand, and great for beginners. It contains hands-on methods and instructions to start you living off-grid or as much off-grid as you wish.

Take the time to decide what you think it means to be off-grid now; you might just get a pleasant surprise or two in these pages.

You will learn what the steps to live off-grid are.

You'll gain an understanding of what it takes, the pros and cons.

When finished reading, you will know:

1. What you need to get started.

2. How to create a homestead layout.

3. The style of dwelling you want.

4. The water and sewer choices.

5. How to heat and use energy.

6. Gardening techniques.

7. How to raise livestock (and why you want to start with chickens!)

8. Ensuring you preserve the food.

9. How to make money from your home.

Numerous topics go into living on a homestead and not depending on anyone else for your survival. Whether you take up hunting or live a more pacifist life by eating only vegetables and fruits, this book can make sure it happens.

Now, with the world in upheaval, hundreds of thousands of deaths from COVID-19, and potential issues of buying enough food from a supermarket, it's time to learn how you can have a sustainable life off-grid. You'll even gain skills that you can make money from when you are ready to employ them.

Chapter 1: What is Living Off the Grid?

There are as many definitions of off-grid living as there are resources available online in the form of homesteading, do-it-yourself, prepping, and green living websites, each differing in the technicalities, but there's one constant in all of those definitions—it's a lifestyle in which you don't rely on everyday utilities such as electricity, gas, and water.

Let's look at the definitions to get a better idea of what living like this entails.

Living off-grid is a lifestyle and system that is independent of remote infrastructures such as gas and electricity. It mentions that all off-grid houses are autonomous, as they do not depend or form on sewers, municipal water, and other utilities.

MacMillan Dictionary defines it as avoiding the commodities, including water, electricity, and gas.

Urban Dictionary (yes, I went there) states something a bit more precise describing it as "Living in an unrecorded and untraceable area through regular means." If you want an example, Ron Swanson from Parks and Recreation is a good one. Urban Dictionary further jests this way of living is a term for a person who uses no form of social media. Yeah, right!

Seeing the definitions, we can theorize that living off-grid refers to a lifestyle with a house completely free from public utilities and employs sustainable living by using renewable energy.

Let's now consider all these points one by one.

Being Independent of Public Utilities

What are the public utilities you can think of off the top of your head? Municipal water, sewers, and natural gas are what we're talking about.

Being independent of municipal water means you aren't connected to the town water and have your own off-grid water system instead. We will discuss more on that later.

Being independent of sewage involves having your field bed, a septic tank, an outhouse, and a propane/composting off-grid washroom.

Last, being free of natural gas saves you a ton in terms of bills, which leaves you money you can spend on having your wood stove, firewood, and other necessities you can use to cook food and heat your house with.

If being completely self-reliant and independent sounds too good to be true, and if it gets you excited, then choosing this lifestyle can be the right choice for you. Not just yet, you must take baby steps. The first question to ask yourself is how off the grid do you want to be, and if you're a person with a family, what do they think of this? This style of living doesn't appeal to everyone.

Being Independent of Electricity

This involves you being independent of an electrical power grid, which you can accomplish by either living without electricity or by developing your systems, such as using solar power, a generator or even turbines that run on wind. That's where the term "off-grid" originates; *the grid is the electrical network.*

Levels of Living Off-Grid

The amount of self-sufficiency you want to integrate into your life is yours to decide. This means different things to different people. There are varying levels of self-sufficiency. Certain people believe that you aren't following this way of life if you still must rely on propane and diesel as fuel for your generator or for heating your house. But others believe that they are merely by installing solar panels in their house in the suburbs. The latter can even sell their generated power to the local network.

Are You Still Off-Grid if You're Still Using Fuel?

Yes. As a hypothetical, let's consider that you're living in a cabin in the woods, heated by firewood, have an outhouse for toiletry, and have your water source. That counts as off the grid. You can use solar power and wind power to generate electricity, in which case you won't be using diesel, oil, or propane. That's considered being completely off-grid, but if you use fuel, that's you being off-grid *only to a certain degree.* Remember, we talked about varying degrees of this lifestyle. The definition is up to you to produce.

Using Solar Power Isn't the Criteria

Technically, if you have a house in the suburbs and are using solar power as your primary source of electricity, you're still connected to the electrical grid. So, doesn't that beat the purpose of living off-grid? In most American states and Canadian provinces, you have the option of selling your solar power, a process that's known as *net metering.*

Are You Allowed to Use the Internet?

Totally. Where you're getting good cell service, you can use the data on your phone to access the Internet. You can have a wireless hub, which allows you to stay connected even if you're residing in a very remote location. Internet access is unbelievably valuable to folk who want to live this way, but still want to generate money by, say, freelancing to sustain financially. And you can use satellite Internet or sometimes visit the nearest town to access public Wi-Fi such as a coffee shop or a fast-food joint. Ask yourself what you want to do with the Internet access and then plan accordingly.

What Does Live Off-Grid Mean to You?

The definition—after including the barebones—is up to you to decide. Early in the process, decide how, what, why and where you plan to get off-grid so you can make plans and strategies to accomplish your goal. For many people, living off-grid is a dream come true, and for others, it's a way to escape the hustle-bustle of living a hectic life in the city. Do you wish to be more self-sufficient? Are you yearning for independence from the power grid? Are you doing it to enjoy nature more? Do you intend to pursue this because it suits your personality, and you consider it quite a lucrative lifestyle? Make up a list of points as to why you crave to do it and what's in it for you. For example, what are the pros and cons of this lifestyle, how rewarding is it, and how tough can it be, what should you do if an emergency occurs, and last, are you cut out for this?

Why Do You Want to Live Off-Grid?

Everyone has their reasons, and I'm sure that you have them too. With those in mind, let's look extensively at the general why's and see if they resonate with your reasons.

1. It makes you self-sufficient.

2. This form of living is sustainable.

3. You'll be relying on renewable energy.

4. This is a more environmentally responsible way to live life.

5. Off-grid living is more practical since we're reusing and recycling and managing our resources.

6. Since you're using fewer resources, you'll leave be leaving less waste.

7. This lifestyle means you get to admire and appreciate nature in a way that's simply impossible in the city. You're more attuned to nature, which will improve your lifestyle, making it healthier.

8. You'll be actively doing chores around the house, working, and striving to sustain yourself, which will make you healthier. The physical benefits are way too many to count.

9. It's a rewarding lifestyle.

10. It will make you happy and content because of more activity and less stress.

11. It will attune you to your roots. Our ancestors used to live this way thousands of years ago. By connecting with your roots, enjoying nature, and utilizing natural resources, you'll find internal satisfaction.

12. You'll be passing that knowledge to the next generation: your kids, students, and younger people. You can share your journey online on YouTube or in a blog, sharing your experiences with people online, and as we mentioned earlier, yes, you can use the Internet and still be off-grid.

13. It is an excellent way to avoid consumerism.

14. You'll be leading other people by your example. Maybe they'll follow and live off-grid after they see the positive impact it has had on your life.

15. The food you'll harvest and prepare will be healthier, as there will be no pesticides or other chemicals used in growing it.

16. The air out in nature is fresher and more exhilarating than the city's air.

17. You must build stuff for yourself including tables, chairs, stoves, makeshift futons, and that sort of thing. This will not only add to your carpentry skills, but will also make you more active, because making things yourself will be a strenuous task.

18. Independence. You'll be more independent of everything.

19. It shall make you more financially responsible.

20. It's a much more socially responsible way to live life. Plus, the lifestyle is safer as compared to living in a city.

21. When you have your homestead, you'll be creating resources for yourself and preparing your own food. Once you make more resources than you need, you'll sell them or sharing them with your local community, which will not only help you earn money but also contribute to your local neighborhood. You get to give back. You can even help local restaurants and eateries by selling them your organic vegetables and fruits. If you're more inclined toward charities, you can contribute to the local food bank.

22. Your backyard will be the entire forest, hillside, and wilderness, which allows you to take long walks, hikes, treks, and explore your surroundings. You'll have more freedom to keep pets. How about dogs? Imagine trekking with them close at your heels.

23. Out in nature, it's peaceful, quiet, and calm.

24. As an off-gridder, you'll always have something to do, whether it is recreational, or something related to your work. You'll mostly be busy, which is an excellent way to pass the time and improve yourself.

25. Living off-grid is taking a long vacation without leaving your home. Just think about all that nature, all that flora, fauna, green trees, placid lakes, and mysterious paths leading into the untamed wilderness. Doesn't that just fill your heart with excitement?

26.

What's In It for You?

1. You'll Get to Escape the System

Even though it's a bit of a political statement to move out into the wilderness and to make you aloof from the "system," once you are out there, you'll feel freer because you'll be rid of debt slavery, corruption, greed, and materialism. If it sounds a little too good to be true, it's because it is. You'll be getting a sense of independence once you aren't shackled by quite the oppressive and intrusive system.

2. You Will Learn Preparedness and Survival Skills

This way of life makes you prepared for many things going awry, especially if you take Murphy's Law into account. You'll be prepared for the worst, honing your survival skills as you do so. This is called prepping.

3. You'll Get to Live a Sustainable Life

You'll reduce the number of resources you consume, thereby making your life self-sufficient and sustainable. You'll be consuming what you produce and will only have room for dire necessities. No excess. Your life will be more balanced.

Two Examples of Off-Grid Living

1. The Cheap Way

Others call it roughing it; I call it the cheap way. It's because this method is not only cheap but also allows you to be completely off-grid. You'll build a small house for yourself, such as a dry cabin. There's no water or electricity in that house; it's essentially just a structure. You'll have to build an outhouse and create a garden for your food, the latter of which is called homesteading. Since this is a cheaper option, know that you'll be relying on one generator for your power. Most days, you'll spend time by the fireplace, the stove, and you will be using candles for light. You'll be expected to get your water, as you won't be near a well or a source. You can either collect rainwater or go into town now and then to buy water. You can also build an underground bladder to store water, but this will cost you more money than this example allows. This form of living is called the Alaskan style. You will be washing all your clothes by hand. Going to the laundromat is not recommended, as it is counterintuitive to your lifestyle. Another thing that you must consider in this style is how to take a bath since you have no running water. You're not going to have an electric stove, so you'll be using more primitive alternatives, and the same goes for refrigerating your food. If you're considering this form of living, consider the pros and cons first, which we shall discuss later, first in an overview, then in a detailed manner.

2. Half and Half

Also known as small-scale homesteading, the half and half style has you depending on the system, but not completely so. You can stay on the grid in this style. Half of your life will rely on the system (power, water, gas, sewage), and the other half will involve you homesteading while trying your best practicing this lifestyle. You'll still have access to the town you're living in. For instance, you'll get to cook your food on an electric stove, wash your clothes in a washing machine, etc. This form of being off-grid is recommended for those of you just starting.

Consider it as tapering off the grid, one step at a time until you're ready to dive in completely.

The advantages and disadvantages of living off-grid:

Advantages

The biggest one is that you are self-sufficient. Consider this: the power goes off in town or your municipality. Thankfully, you won't be affected. The same goes for any emergency that takes place at, say, the water plant, for example.

You won't be spending exorbitant amounts of money on your bills.

Last, freedom is the biggest advantage, and you can't put a price on that.

Disadvantages

You must move a lot because most places won't allow you to be off-grid. Your resources will be scarce if you move to a place that's not close to town.

You must consider the impact it will have on your family.

It could mean legal trouble. Before you try this way of living, consider the local laws.

Your life will be drastically affected, such as your inability to run a shower, iron your clothes, and wash clothes and utensils.

The Rewards and Challenges

1. You must source your food. Your priority will be to secure your food sources. This entails getting pesticide-free produce, establishing a secure source, and arranging meat that's humanly raised. This will require a great deal of challenging work on your end because you must set up the whole system and maintain it, harvest food, and butcher the animals. Last, once you have all the raw materials, you must be

careful that they don't go to waste, but remember that this will pay off monumentally.

2. You will have to build your own home. Before delving into this venture, you must brush up on your home building skills and make sure that everything you build is durable. In return, you will get complete freedom for choosing the materials, and you can always get non-toxic stuff and renewable resources for your home, thus positively contributing to the environment.

3. Nothing gets done if you don't do it, so if you're the type of person who lives a procrastinating and passive lifestyle, this honestly might not be the right fit for you. You must take into consideration that you will have to chop your wood, grow your food, and empty your toilet. What do you get in return? Freedom, yet again. You're not going to have to report to a superior.

4. You won't be at the mercy of the grid, which will allow you to weather all the storms that come your way. You're going to have to work a lot, sure, but you'll also get consistency, stability, and hardiness in your character.

5. People will not get you. They'll think of you as eccentric, viewing your lifestyle with apprehension, curiosity, and skepticism. It will also be a conversational hurdle, as you won't be able to relate to people, nor will they be able to relate to you.

How to Get Started Living Off the Grid

1. Get a Wood Stove

Installing a wood stove in your home will heat your water, your food, and allow you to dry your clothes even when it's raining. It will also save you a fortune in terms of energy bills.

2. Learn From the Pros

There are a ton of guides (including the one you're reading) online that can help you get motivated on how to get started.

3. Learn How to do Things by Hand

This is key. You'll be doing most of the work by hand since there won't be all that many convenient devices at your disposal. A few examples include kneading the dough, using a manual coffee grinder, and using a handsaw.

4. Prioritize Your Evenings

Now that you've decided to try this lifestyle, understand that your evenings won't be spent in front of the computer or the television. You can use that free time to learn skills such as cooking, sewing, whittling, and knitting.

5. Pick Out a Location

If you're serious about going off-grid, you should start looking for a piece of land where you can build your homestead. It's not an easy task to find affordable acreage far away from the city that has all the commodities and amenities nearby if you should need them. It's going to require thoroughly researching real estate, land prices, and location.

Chapter 2: Are You Cut Out for Off-Grid Living?

Are You Aware of the Challenges of Living Off the Grid?

Because let's be honest, as popular and appealing off-grid living is, it comes with its own sets of challenges and considerations. Know beforehand if this is something you're cut out for or not. Before jumping right into this lifestyle, let's consider a few of the challenges in detail.

An important statistic to consider here is that upwards of 200,000 Americans now live off-grid.

There's a steep learning curve to this way of living, but once you're past that curve, things will start to fall in rhythm.

1. The Location

Location does not just pertain to people who live in the city. As someone working to live an off-grid life, you must consider that the land you pick is just about rural enough that there are no utilities available, such as electricity and gas. Now, second, consider the land that you need for your homestead. What are the essentials you need to sustain yourself? Are you planning on raising livestock? If so, you must get land that offers shelter, stores their food, and allows them to graze. You will also need a water source such as a well. Last, you'll need space and land for your waste management, recycling system, and your garden.

There's also the issue that many municipalities don't allow you to dwell as an off-gridder, requiring that you hook your home up with the power grid. In certain cases (in a few cities), they'll allow you to go off-grid with the added caveat you supplement your power with alternatives and don't cut off completely. All this calls for buying land that's nowhere near a city or a municipality.

2. The Power Source

The next important bit to consider is the power source. Yes, you can go into *ye olde* times by cutting yourself off completely and not relying on any power sources, but if you do decide to have a power source, it'll make your life a bit easier. The easiest route here is getting solar panels. It's a long-term investment with a decent ROI (return on investment). Another option is a generator. There are a dozen alternative sources at your disposal, and they all depend on what's your budget, what's your location, and how much power you need. For a starting point, check your electricity bill and see how much energy you are using. Now, plan accordingly.

Once you've set up your power, you will quickly realize that you are consuming more energy than needed. Consider this as a withdrawal symptom from your days living on the grid. Conserving energy will be beneficial here. Try running only one appliance at a time and do that when it's daytime so you can get the most out of your power source if it is solar energy. Don't rely all that much on your energy source, though. You are, going off-grid. Ease yourself in with alternative methods, such as using a stove to heat your food and home and using gas instead of an electric stove.

3. The Food

As an off-gridder, you'll be sourcing your food. This means growing everything you need to eat, raising cattle and livestock, and getting a garden established. The kitchen garden you establish in your homestead must expand gradually, as you will start producing more food than you need so you can sell it and generate income to supplement your lifestyle.

Possibly, the biggest issue you'll face in terms of food is a balanced diet. You need to have variety in your food since it's just impossible for everyone to raise livestock at the start of their homesteading. So, what do they rely on in the initial phase? Veggies. That's right. You can produce vegetables in your garden.

Another challenge that you'll be facing is cooking and preserving food. Both these skills are essential in homesteading and require diligence for the offseason in which your garden isn't producing as much.

Many homesteaders go the hunting route. If you're the sort who likes to hunt, this will be an excellent opportunity for you to hunt game, skin them using a tactical knife, butcher them, prepare them, eat them, and preserve them.

4. Water Supply

You must have a water supply nearby. It should be a reliable water supply and one you can depend on when things go awry. Trust me, things can go awry fast when living on your own in a rural setting with nothing to rely on but yourself. Add to that the unavailability of water, and you'll have a complex situation with no harvest, no cattle, and nothing to sustain you. Some lands are located near freshwater sources such as rivers, but they are rare to find.

Instead, bore a well and use a hand pump to draw the water out whenever needed.

Another important thing to consider when making the shift from your "normal" lifestyle to an off-grid lifestyle is that you shall miss running water. It's a commodity that we take for granted, and when it's not available to us, such as in the case of being off the grid, we'll feel out of place. A workaround includes getting a generator so you have running water at hand, but one factor about generators is that you need to keep buying diesel for them. Is this something you see yourself doing, or would you rather save costs on fuel?

Last, you must keep the legal aspect of it all in mind. Are you allowed to use the water on the land that you've bought? Usually you can use the water in your property for your personal use with no issue, but, say, if you are using water for livestock and farming, keep in mind the laws of the municipality. Consult a lawyer beforehand, if necessary.

5. Time Management

Think of living this way as having multiple jobs all simultaneously. From harvesting food to cleaning the outhouse, tasks are abundant at hand. You're going to have to manage your time expertly so that everything goes forward smoothly. There's no room for procrastination here. If you're sowing seeds, they should be sown in the correct season. If you have animals, you must feed them regularly and milk them on schedule. To protect your homestead, you must set up fences. To store your harvest and food, you must build shelters.

All that work will keep you more than occupied and on your toes. It gets tiring and gets jarring at times too, but overall, once you've picked up a rhythm, you'll thank your past self for it. What's the secret to it all? Time management. You can keep a journal or a to-do list with you. If you still have a phone, you can store your schedule in that.

To find time to do all the above-mentioned stuff and more, you must become an early riser. Also plan out the next few months or a year for long-term projects. Besides keeping track of your time, also keep track of the weather patterns.

6. Budget

Let's get one thing straight—this is not a cheaper option. That false image is the romanticized stuff fit for movies and books, where the protagonist moves out of the burbs and heads into the wilderness where he finds internal peace. The last part is not exaggerated. You'll get your fair share of internal peace, but it will come at a cost. There will be a slew of upfront investments that you must make to succeed. Also, you shall have to keep up with the present rates for livestock and their feed, and the cost of growing your crops.

The most important thing where a considerable amount of your budget will go is the power source. Setting it up will require you to make a huge investment in solar panels or generators. If we're realistic here, this will cost you tens of thousands of dollars, especially considering that you will have to set up between 15 to 30 solar panels to support the needs of your homestead.

If that sounds like too huge of investment, start with baby steps. First, grow your food and cooking it, then move on to reducing your carbon footprint, and then become more conscious of your power consumption. By taking these measures, you'll end up not only saving money for your new lifestyle but also get in the essential habits needed to survive.

7. Isolation

Let's consider the hypothetical protagonist who found inner peace by heading off into the wilderness and cutting all ties with society. That's not always going to be the case. A lack of human interaction will serve as a challenge to you if you're a social and extroverted person. Yes, most homesteaders do start their new lifestyle to escape from the city and its hectic life, but it need not be a solo venture. Contact other like-minded people and form an off-grid community where you can go whenever you feel alone and isolated.

If there isn't a community nearby you don't fret. There are dozens of communities online where veteran off-gridders share their advice with newbies, and people share their tips, tricks and experiences. Consider joining them.

Are You Ready for an Off-Grid Lifestyle?

1. Your Mindset

You should have a conservation mindset to pursue this lifestyle. In our regular homes, we become accustomed to a lifestyle of excess. Running water is available whenever we want it for washing, laundry, and taking baths. There's always electricity to rely on for our gadgets, for changing the temperature of our rooms, and for running practically everything ranging from stoves to garage doors. You and I pay our bills every month, never once considering where all our resources are going and where they come from.

There's a need for a radical shift in your mindset when you decide to go off-grid, especially in terms of resources because they are limited and deplete the more you rely on them. How does one avoid that? It's simple:

- At a given time, use only one appliance. Say, at breakfast, you need to use the blender, toaster, and the coffeemaker; you will have to use them one at a time instead of using them simultaneously.

- After you have used a charger, unplug it.

- Use power-heavy appliances such as washing machines and dishwashers in the daytime, particularly if you're using solar power.

- Take baths only in the afternoon when the water is at its hottest because of the sun.

- If you can cook with propane, all the better.

- Your priority for heating your house should be a woodstove rather than any electrical heater. There are no thermostats out here.

- Open windows as often as you can for ventilation and nighttime cooling.

2. A Design Specific to the Site

The architecture of the house is critical when considering going off-grid, or even just trying to be super-efficient in your regular home.

Remember, you must consider the building's passive solar design—as in placing windows and skylights to maximize sunlight or exposure to the southern side—and the orientation of the building. You'll be getting free heat and light out of the deal if you plan your house accordingly.

Think about having a super-tight enveloping building to help you manage energy loss, particularly when there's an enormous difference between the external and internal temperatures.

Everything that you do should be energy conservation centric.

3. Power Generation

An average American home utilizes around 10,000 kilowatt-hours per year. Before jumping the gun, first, analyze your bills for a pattern and see how many kilowatt-hours you are utilizing on a monthly and annual basis. Then consult with your architect and determine the shift in your lifestyle and how that can improve your power costs.

Think about all the appliances you use every day, such as your irons, dehumidifiers, hairdryers, and hair straighteners. Have you considered how deeply how you are tied to the grid and how invasive it is in your everyday life?

How are you planning to power your homestead? Will you be using solar power, geothermal, or wind? Regardless of the source, you need to have backup power in the form of a generator.

4. Water Collection

In terms of essentials, a renewable water supply is important. A common way to go about it involves a well or using freshwater. After you have established a water source, pick out a plan as to how you will get it to your home. Will you use pumps and a filtration system?

Once that water is at your home, you'll need a holding tank for it. If you're in an area that gets heavy rainfall, you can set up rain barrels to supplement your water consumption. This is gray water, not recommended for drinking use. Instead, use it for your landscaping needs.

5. Waste Disposal

This bit is a little complicated as you're both bound and unbound from the government simultaneously. Yes, you are still (hypothetically) off-grid, and normal building codes are not applied here, but there are entities such as the Environmental Protection Agency and other health departments that require you to dispose of your waste safely according to their regulations.

One way to go about it is by installing your septic and drain-field system that will return all waste into the ground, where it will all be utilized in nutrients. Yes, the system will take up a lot of space when you're installing it, but once it is all done and dusted, it will disappear. Last, you will need to maintain both these systems.

Is Living Off the Grid the Right Choice for You?

Understand that this way of living isn't for everyone. Not everyone is cut out to tough it out in the wilderness. Although, if you're feeling overwhelmed and thinking to yourself that you might be ready for this, here are a bunch of questions that shall put it into perspective.

1. Do You Value Your Time?

When you're on the grid, you're tied down by a system that constitutes a job, giving time to your television and computer, going to and from a grocery store, managing your social life on top of your work life, and seeing to all the expenditures that come with the territory of living in a city. To say this lifestyle consumes a lot of your time would not be an overstatement.

Once you're living off-grid, most of the time you get free from your homesteading tasks will be yours. You can even quit your job and earn money by selling your organic produce. You're not going to spend time in front of the computer, you won't have to keep on checking your phone, and you won't be sitting passively in front of the television. All that free time will be yours to do with as you please. You will be working on your schedule and for your benefit. As grueling as the lifestyle is, it is just as rewarding.

2. Your Friends and Family's Reaction?

This will be a fantastic opportunity to weed out people who aren't your real friends. When you tell them about your decision to live this way, those who care for you'll support your choice and tell you to go for it, whereas those who don't have the best intentions for you in their heart will tell you many reasons you should not do this. A few of your friends and family members will be inspired by your example, others will be jealous, and the rest will be indifferent. It's up to you and you alone to decide to be off-grid or not.

3. Will You Have to Quit Your Job?

Once again, it is up to you. If you are planning on being a full-time off-gridder, then, yes, quit your job and focus solely on your new lifestyle. If, on the other hand, your job is essential, then consider the half-and-half method we discussed, in which you'll be off-grid but stay in the city so you can commute to and from your job. Then it is up to you to decide. If you have a family, then consider their wants and needs, ask for their opinion, and then produce a system that can sustain you.

4. Can You Afford It?

Start saving money the moment you go off-grid. You don't have to have all the money right away. You can do this in easily manageable batches that won't be heavy on your wallet.

5. What to do with the Money You Save

The saved money will go to all the costs such as energy, water, waste management, and food.

6. Do You Know What You're Doing?

Kudos to you if you know what you're doing, but if you're unsure about what steps to take, read up on a couple of resources, consult online, and watch as many videos on YouTube as you can. Asking veteran off-gridders for advice is another way of making sure that you're doing the right thing.

7. Do You Have Contingencies?

Again, with Murphy's Law being ever-present, all that can go wrong will go wrong, so you need to be mentally – and resourcefully – ready for that. Write it in your journal. Keep track of all the things that can go south and produce a series of steps to tackle them.

Chapter 3: The Pros and Cons of Off-Grid Living

We discussed the advantages and disadvantages briefly in the previous chapter. Let's take a more detailed look at them now.

Pros

1.Peace and Quiet

It's utterly quiet and calm when you start living off-grid. Gone are the sounds of traffic horns and people cursing at each other in traffic jams. You are trading in the overwhelming sound stimuli for a feeling of peace and quiet that will leave you feeling peaceful. You won't have a constant bombardment of noise after noise, plus all the other forms of pollution that add stress and anxiety to your life, such as all the many screens you are used to watching—phone screens, TV screens, and LED advertising panels. Because of this lack of incessant stress and anxiety, you will be less overwhelmed in life. You'll be getting a digital detox when you move off-grid. At first, you will have your fair share of withdrawal symptoms, but once you're attuned to your new lifestyle, you shall appreciate it more for all the perks it grants you.

2.Quantum of Solace

If you are planning to go off-grid all by yourself, it will minimize your lifestyle in a whole new way. You won't be interacting with all the people that were always around you. This includes relatives, friends, colleagues, and neighbors. You shall be living on your own, having to answer to no one. Initially, you shall feel a bit out of place, sure, but as you get into your new lifestyle, you shall appreciate cutting off all (well, most) ties with the world. Your life will have fewer distractions in it. It will be a less intrusive life, one where everyone won't be bothering you all the time, which translates to less stress.

3.Land Security

Doesn't every one of us want to own our place, a place that we can call home? Well, here's your opportunity to do so. Not only is the property in rural areas cheaper, but there is also a lot of it. Once you buy your land, you will not have to worry about things like rent and mortgages. You'll be in total control and ownership of your piece of land and can do whatever you want to do there, such as building new extensions, keeping livestock, expanding your homestead, setting up your garden, and so on.

4.Surrounded by Nature

This is probably the best advantage of moving off-grid. Greenery and nature are scarce in cityscapes, limited to parks and small gardens and the odd tree here and there. When you adopt an off-grid lifestyle, you find yourself pleasantly surprised by all the surrounding nature such as: green grass, lush trees, babbling brooks, peaceful lakes you can swim or boat in, and the clear blue sky unadulterated by the fumes of vehicles and factories. You'll become healthier in just one month in your new lifestyle and will take long walks to appreciate nature more. This will get you in touch with your roots. It will also make you less anxious and depressed. All this nature around you will be very therapeutic and inspirational.

5.Time

Time tends to stand still in the wilderness. Granted, you will be busy managing all the aspects of your new lifestyle, but all the time you get free from that will be yours and yours alone to do with as you please. There will be no interruptions. You will get to decide your schedule for the day, answering to no one. You can spend this time— as we discussed earlier—in learning new skills and developing new hobbies. You'll be able to do whatever you want, given you go full-on off-grid. Doesn't that sound swell?

6.Opportunity for Introspection

Once off-grid, you shall have an abundance of time and energy to explore all the ideas in your head, once and for all. This includes everything from your daily thoughts to plans about starting a new hobby or finally getting that book you promised yourself you'd write. Learning a new skill goes hand in hand with off-gridding. You'll be able to introspect and discover what sort of person you are, what you inherently like and dislike, and where do you want your life to go from this point forward. All this introspection will grant you more insight into your life. And the best part about it is that you have time to meditate.

7.Fresh Air

Do you know how important it is to breathe fresh air? If not, then consider all the thousands of vehicles and factories that contribute to the pollution inside the city. Essentially, what you're doing is slowly suffocating yourself in the city. Once out in the open surrounded by fresh air, you shall feel more energetic, cleansed, and healthier. When it's nighttime, you can see all the stars, and when it's daytime, your view of the sky will be pristine.

8.Mindful Living

Meditation and mindfulness are critical to personal development and living a healthier lifestyle. Once you're out in the open, you can set up a schedule for when, where and how long to meditate.

Declutter your mind. This will catalyze the simplicity in your life since all the choices that you will make from this point forward will be deliberate and yours to make.

9.Limited Access

Living off-grid will limit your access to the closest town and technology, making you more self-reliant and less dependent on systems and tech. All this adds to more time on your hands. This is only if you decide to go completely off-grid.

10.Social Life

If you are an introvert and living a social lifestyle drains you of your energy, the disconnection from people and social responsibilities will allow you to live life on your terms without the pressure of meeting and staying in touch with people.

Cons

1. Start-Up Costs

Everything involving off-gridding comes at a price. Unfortunately, that price is a bit too steep for most people. This is sometimes the only factor that deters most people from making the shift. Purchasing all the equipment, getting it installed, and maintaining it is awfully expensive. You're going to have to start saving up at least a year or two before you can go completely off-grid. It isn't just the start-up costs you must consider; there are dozens of factors at play here that will all require you to have deep pockets. But this should not be a deal-breaker for you, since you can start off-gridding in batches, breaking it all down into smaller modules that are less taxing on your money. Remember to take one step at a time.

2. Maintaining Everything

If you're a one-man (or woman) off-gridder, you'll find yourself busy with maintaining everything round the clock. As we mentioned before, nothing gets done if you don't do it. Whether it's renovations around the house or repairing the worn-out fences surrounding your property, maintaining everything will be not only tiring but also expensive. There will be days when you question whether it is worth it. You'll have to remind yourself the reason why you started off-grid living in the first place, and that it is indeed worth it.

3. Waste of Energy

All the energy you won't be utilizing will go to waste, and here's the thing—you will not be conscious of your energy consumption all the time. Think of it like this: are you conscious of your energy consumption at home? It takes more than a couple of months to get into the conservation mind-set where you begin to be more conscious of your resources, but for the first few months, you will end up wasting all the energy you won't need. This is so because surplus energy cannot be stored anywhere. A workaround to that is using only those necessities when you need them, then unplugging them.

4. Difficulty Communicating

As a full-time off-gridder, you'll cannot stay as connected with people from your previous life. If you intend to go completely off-grid—as in no computer, no phone, and no Internet—the prospect may sound appealing at first, but once you are doing it, you will realize that it's difficult to not communicate with people. Also, if emergencies occurs, where will you go, or to whom will you turn? You cannot call people when you want and vice versa. Honestly, this is the biggest downside of living off-grid. You'll be disconnected from your friends and family and only be able to meet them when you make one of your trips in town to get supplies and equipment. Are you ready for that?

5. Disconnect From Technology

There are degrees to which you decide to live off-grid. In many cases, you will not disconnect from technology and will have your phone and television and computer at your home, but where you're going full-blown off-grid, the disconnect will come at its costs. You won't be able to keep up with the news, won't be able to work remotely, and won't be in touch with your friends and family. This is why most off-gridders don't go completely tech-free.

6. Limited Social Activity

Since you're living off-grid, you won't be able to attend parties, social gatherings, and events in your town since most of your time will go towards maintaining your lifestyle and tending to your homestead. Besides, the commute to and from your home will be too long and not worth it.

7. Uncontrollable Factors

We have discussed this previously. You must be prepared for the worst-case scenario, which is that everything that can and might go wrong. Sometimes something will happen you weren't mentally prepared for, and you won't know how to deal with it, despite the contingencies you have put in place.

8. Too Much of a Good Thing

Spending too much time all by yourself will give you a case of cabin fever, or at the very least, it will make you quite antisocial. To compensate for your lack of social life, you will throw yourself in your work, which will drain you, causing a burnout.

9. Money Problems

We're not just talking about the start-up costs. If you're not working a full-time job, things will get a bit tough for you when you must spend money on new items for your homestead. From where will you generate your income? Part-time work? Remote work? Have you considered this aspect thoroughly?

Chapter 4: What You Need to Get Started

As with everything, getting started is the hardest part of off-gridding. It's quite a drastic shift in your life, isn't it? The mere notion is enough to make you reconsider something as daunting as going off and living in a remote wilderness. The number of things you must do can also get quite overwhelming. Where do you begin? Do you start by learning woodworking and carpentry, or do you learn how to preserve your food?

To make this shift less scary, let's produce a checklist that shall modularize the whole thing into manageable steps.

The Initial Planning Phases

In this phase, you should, first, get all your financial affairs in order. Let's begin by making a budget that you'll spend on your new lifestyle, then moving on to reducing all unneeded expenses, such as buying new clothes, eating at fancy restaurants, and paying your Netflix monthly bill. Once you have done that, you'll notice that living off-grid doesn't seem as difficult anymore. Good. This is why this is the first step. Second, you should save money from this very moment. If you have any debts, pay them all off before going off-grid.

Once the budget part is out of the way, research on which off-grid style serves you best. There's a wide variety of styles, two of which we discussed in the first chapter, and the rest of which we will discuss later. It's all about finding the right balance between going off-grid and staying somewhat in touch with your old life. You're going to have to keep up with your relatives, after all. Sooner or later, you will also need tech to rely on. So, doing research about which style suits you best is important.

After that, start researching the local laws about off-grid living. Different states have different laws so consult with a lawyer beforehand, and if the laws don't allow you to live off-grid, it might be an excellent choice to move away to another state or even another country. You're going off-grid. It doesn't matter where you are, does it?

Now that you've researched the lifestyles and done your budgeting and decided which style suits you best, decide on how far you want to go off-grid. Do you want to go completely primitive, or do you want to stay in touch with technology? What elements are you willing to incorporate, and what elements are you willing to let go? Can you live without electricity, sewage and water? What are your workarounds for that? Think about it all before you take the plunge into your new lifestyle.

If you have health issues, consult with your doctor, and ask them how your health will factor into off-grid living and how frequently should you visit the hospital for regular checkups.

Ask yourself again, why are you going off-grid? Are your reasons strong enough to warrant a tectonic shift in your lifestyle? Are you motivated enough? What's holding you back from going off-grid? What are you going to do once you've successfully gone off-grid?

List all the skills you already have, such as woodworking, gardening, carpentry, and first-aid. Now that we have those out of the way, we can focus on the skills you don't have and need to learn to live off-grid, but that comes later. Remember the importance of baby steps.

Last, figure out a way to earn money once you've moved off-grid. Are you going to sell the produce from your garden? Do you intend to become a freelancer? Will you commute to and from work every day? It's completely up to you, but you must get it out of the way in the initial phase of your planning.

Learn How to Produce Own Your Food

First, gardening is an important thing you should learn about your off-grid lifestyle. Read about it online; follow YouTube channels that teach gardening and then start applying your knowledge in your current home. For starts, pick a few plants and plant them in the flowerbed and tend to them for a month to see if you can do it or not. It's not that hard of a skill – you will surely be able to do this! There are various gardening methods to choose from, such as greenhouses, aquaponics, hydroponics, and permaculture. Do your research about them and select one that suits you best while also being the most convenient for you.

Second on your list should be *hunting*. Are you interested in sourcing your food through hunting? If so, the first thing you need is a hunting license, the second thing is the gun itself, the third is practicing with the gun at a shooting gallery, the fourth is learning about animal tracking, and last, how to butcher animals.

If you don't want to hunt and instead want to raise animals for your food needs, visit a farm to see how they're doing it. If you are open to the idea, you can volunteer at that farm to practice raising animals.

Buying the Property

Pick out the area where you want to live, gauge it by checking its accessibility to water, the proximity of the neighbors, soil quality, and available natural resources on your property.

If you're satisfied with the land that you've chosen, go ahead, and meet a real estate agent and buy the land. You'll be surprised how relatively cheap it is. This is because, as compared to urban areas, the rural property doesn't cost as much, which is another reason the off-grid lifestyle is very appealing.

Planning Your Home

The first thing you want to do is consult with an architect or an engineer specializing in green homes or *eco-friendly* homes. It'll do you good to learn about green home building materials and methods. Once you've done that, it's time to decide the size of your home. If you're strapped for cash, consider building a house that is small in the beginning but one that can be added to later – perhaps in modules. Now, decide whether you want to hire people to build the house or if you want to do it yourself. Doing it yourself will teach you a lot of skills that will come in handy later - skills such as carpentry and construction.

Then, plan your property's layout in terms of exposure to the sunlight, water access, the location of water resources (well, lakes, rivers, streams), the gardens, waste disposal areas, and outhouses/outbuildings.

Plan the water supply system and plumbing by researching rainwater harvesting systems, hiring a geologist to locate wells on your property, and determining where you want your latrine/septic tank placed.

Next, we move on to the electric system. The first thing is a no-brainer. Do you still want to be connected to the grid or not? Of course not! Why else are you pursuing this lifestyle? Decide whether you want to use wind power, solar power or a combination of both. Do your research and consult engineers (solar power engineers and/or wind power engineers) about your equipment needs and your energy requirements.

Now, we move onto the heating system. What kind of insulation are you planning on using? What kind of heating system do you want? A wood furnace, a wood stove, solar heating, power generator, animal dung, or coal? Specify it in your checklist. To buy fuel, you should secure access to it by now.

You're also going to have to build a root cellar for the storage of food, particularly root crops.

What are your transportation methods? If you do not know the mechanics of automobiles, I suggest that you brush up on basic mechanic skills, since you will need them at your off-grid setup whenever a vehicle malfunctions. You won't always have a mechanic at hand, and you're striving for independence; it makes sense to learn auto mechanics. Now, consider what kind of vehicles you can use at your property according to the terrain. Examples include bikes, quad bikes, motorbikes, four-wheel drives, and trucks.

It's time to reevaluate your budget so that you can plan for permits, labor, building expenses, machinery, repairs, and other costs. Prioritize your costs. It's recommended that you produce a project plan for your home, itemizing each item and its expenses. Last, consider which projects are essential (they should be your priority) and which projects can be delayed.

If you will work from home, set up a framework for your off-grid office.

Moving into Your Home

The first step to moving into your new home is its construction. Decide when and how you will move in. If it's difficult to move in right away, you can move in at a later time and still give your time to your site by visiting on the weekends and setting up your camp or staying in an RV to ease yourself into this new lifestyle.

Plumbing and water are critical to off-gridding. First, install a toilet system, and then move on to drilling a well. You can set up a rainwater and greywater harvesting system as well as setting up a solar-powered shower to give you that warm bath at the end of a long day.

Install your heating system and stockpile for the winter. That includes firewood and fuel.

Now comes the electricity bit. Install a solar system as your primary source of power and buy a backup generator just in case you need it. Start practicing a reduced energy lifestyle so that when you move in, you're not overwhelmed by the scarcity of power.

Manage your waste by starting a compost pile and learning how to reduce waste and reuse it.

Starting your garden is another crucial step you need to take. Calculate how much food you will need for the winter and start planting accordingly. What kind of food do you want to grow? Make a list and then get started on designing your garden. Again, reevaluate your budget so you can afford tools and seeds. Make sure you plant your crops in the correct seasons. If you want nuts and fruits, you can plant an orchard for it. For your first crop, start by sprouting seeds. Learn how to repair and maintain your gardening tools.

Get chickens, build a coop for them, buy heat lamps, buy chicks, and raise them indoors. Then, move them to their coop when the weather is favorably warm.

Then, get livestock and build pastures, fences, and pens to keep them in. If you need a barn, build one. Visit livestock fairs to get an estimate on costs. How are you going to transport your livestock to your off-grid site? Also learn veterinary care for your animals.

Start Building Your Greenhouse and an Aquaponics System

Keep your health in check by learning first aid and natural remedies, and create/maintain a medicinal garden at your site. To avoid diseases, eat healthy food, stay active, and make regular appointments for annual check-ups, as needed.

Finally, be happy and content knowing that you have started living off-grid and have accomplished a goal that most people only dream of – but never actually pursue. You, on the other hand, have made your dream a reality. That's something to be proud of. You've earned bragging rights; go ahead and flaunt it.

We have only discussed the basics of what you need to get started. Now we will study each aspect in detail.

Chapter 5: Creating a Homestead Layout

After covering what we need to get started, let's get into a few layouts more extensively so you can decide on which homestead arrangement suits you best. We'll be covering around 28 of them to ensure we cover the topic exhaustively. Let's get right into it. For reference to the images and design, we are using this link:

https://morningchores.com/farm-layout/

1. The Tiny Backyard Layout

This is the most suitable option for those of you still living in the suburbs and yet want to be off the grid. In this blueprint, you will be harvesting your food in a back garden or on a small piece of land you own. You will set up a chicken coop in one corner of the place, vegetable boxes in the center, fruit trees by the eastern wall, and herbs by the western wall. A stone pathway can lead to your small backyard layout. The biggest advantage of this design is that you don't have to move to a new place. You can still be off the grid while living in the suburbs.

2. ¼ Acre Layout

This is for those of you who own land in a rural area. How much land? Well, a quarter of an acre. Suppose you're confused about how you will design a blueprint that suits your purpose. It will not only give you space to grow your food but will also have plenty of room for livestock such as chickens, sheep, rabbits, and goats. You can start by dividing the property into around five modules. The topmost area of the land should be a goat pasture for your livestock grazing; below it, add another goat pasture where you can plant fruit trees and a chicken tractor. To the upper left of it, create a goat pen, and below that, a chicken coop. To the utmost left, place vegetable boxes for herbs and plants. Does that sound manageable?

3. ½ to 1 Acre Layout

There's plenty you can accomplish in this amount of area. Not only can you raise your livestock, but you can also plant all the plants you could need, and a little play area for your children.

In just a half to a whole acre of land you can add nut trees, goat and sheep pastures, a chicken coop, fruit trees, pens for your goat and sheep, a compost pile, rabbit and chicken tractors, alfalfa, a tool shed, vegetable boxes, an herb garden, and your house. You'll still have a space left over that you can turn into a calm and serene orchard or garden.

4. 2 to 3 Acres Layout

If you have bought two to three acres of land for your off-grid homestead, it means that you're considering this seriously. Good for you. The biggest advantage of this much land is you can plant oats, wheat and alfalfa while also building pens for pigs, goats, and sheep. In two to three acres, you're also able to build a medium-sized barn for your animals. Besides the barn and the pens, you can add rotating pastures, a garden, a shed, a coop, and an orchard. You'll be able to build a good-sized house in that much land as per your requirements.

5. Urban Permaculture Layout

Permaculture is a design principle that centers on whole systems thinking, directing, and simulating the conditions and features that exist in natural ecosystems. At least that's what Wikipedia says about it. If you're not planning on moving out into the wilderness and intend to go off-grid at your present home, this design will be most beneficial for you, but there's a caveat at play here. You're going to have a lot of land at your disposal. This design is very organized, very structured, and well thought out. There's a décor element to it, as well as functional utilization. You can plant anything you want, from an orchard to a garden, in this design.

6. Edible Garden Layout

If you live in an area with a warmer climate, this design would be a clever idea. You can grow your pomegranate trees, citrus trees, vegetables, seeds, and nuts. The edible garden is a sustainable and functional design.

Here's how it works out: in the center, you have a house. Above it, there's a patio extension. Above that, there's a parking lot with ample space for two cars. To the right of the house, you can set up a garden with figs, avocados and oranges. To the left of the house, there's supposed to be a bed for herbs. Further left, you can grow lemons, oranges, and mandarins. There's a shed beside that, above which you can plant apricots, nectarines, hazels, peaches, and apples. This design is the best option for vegetarians. Speaking of which, put a greenhouse, a wood store, and a toolshed on your land.

7. Designer Micro Layout

This is yet again an awesome design that shall appeal to you if you're a vegetarian or vegan. This design is not only functional but also decorative, with a big house in the middle and plants and trees surrounding it from all sides. Loquat, hawthorns, and ivies can grow on the right side of the house. You can add a quaint deck to the far right, overlooking vegetable crops. The layout's top wall should contain all trees, ranging from pomegranate to pears. Cane berries, Nashi, quinces, apples, holly, apricot, and cherry trees can be planted on one side. To the left, you should have a pond around which you can grow thyme, dwarf lemon, and Mediterranean herbs. The only drawback of this design is that there's no place for livestock—only fruit, and vegetables. Although on the plus side, you have a very ornamental garden at your disposal with a ton of fruits and vegetables available in both on and off-seasons.

8. Designer Chicken Layout

The designer chicken design is optimal for raising chicken in a convenient and neat manner. This homestead style is aesthetic and functional at the same time. You can have a longhouse with a carport on one side, a patio on the other, and two doors—one for the kitchen and the other for the laundry—and all the plants and trees that we've mentioned in the previous sections. You can also add the main cropping area for your crops. If you have pets such as cats or dogs, you can add a fence around their area so they don't interact with your chickens. The chicken shed is on the corner of the homestead, with a butchering station next to it, though not so close that your chickens can see their fellows being butchered.

9. All-Inclusive Layout

Are you ready for the big one? Then look no further. In this design, you will have everything, and by everything, I mean *literally everything* your off-grid homestead requires, from sheds to nurseries, from plants to pens, from trees to tracks. *Everything.* Of course, it shall require just as much space, but in return, it will make you completely self-sufficient. Not only is there a placement detail for each and every single element of the homestead, but there's also an aesthetic element to all the placements. It also includes a design for your own power and water system.

10. Basic Farm Layout

The basic farm layout has everything that a farm constitutes, from barns to stables and silos. There can be cabins you can use as a granary and a woodshed. You can dig for a pond you can use as your water source in dire situations later. Essentially, this design is perfect for you to emulate a farming lifestyle. There's just the added requirement of a large amount of land for this form of design, as it can't fit in less than an acre, but you can take the layout and remove things from it to suit your purpose, thus accomplishing it in a smaller area.

11. The 1-Acre Dream Layout

It's a dream layout because it is completely independent of the amount of area you have. You can modularize it into smaller chunks and fit everything well within an acre. The homestead will be functional and compact, fulfilling your every need without making you exert yourself by traveling all over your property. It's manageable, dependable, and sustainable. You can plant vegetables in one area, crops in the other, build a coop for chickens, a shed and pens for livestock, an orchard for fruits, and a greenhouse. It depends on what you want instead of the other way around. You'll also have your choice of home styles to pick from, which we'll discuss at length in the next chapter.

12. Permaculture Layout

The previous permaculture layout we discussed was an urban one. This is different because it requires a lot of area for your permaculture. This is your best option to live in a manufactured home as opposed to, say, a cottage or a hut. You can plan to build this design all-around your house. This form of layout is modeled after an old-style farmhouse. It has versatile features and maintains an aesthetic sense exclusive to permaculture farming. You can also use your plants as natural hedging for your properties, which is an effortless way to set up boundaries.

13. 1/10th Acre Layout

It's a simple, elegant layout with seven main components in a grid-like setting. The first section will be on the top left for vegetables with eight beds each with each bed being 4 x 8 feet. The second section will have fruits and nuts in it, including multiple fruit trees, vines, berry canes, and strawberry beds, all of which will be along the fence line. Third, there shall be a bed for herbs beside your home. There isn't a lot of space for grains, so we're excluding them from the list. You can have a smalltime poultry pen with six chickens in it, and you can keep rabbits as well since there isn't room for larger livestock here. In terms of wild food, you can add a beehive or two for honey.

14. The Garden Layout

This layout is not exactly a homestead layout, but consider it if you want to add a nice garden to your existing house. That way, you will become self-sufficient in terms of food, if not other aspects such as electricity and water. The way to sufficiency is gardening and livestock. We'll discuss the power and water part in detail in the upcoming chapters.

Having a garden layout will provide you with all the food you could need, giving you your daily dose of minerals and vitamins, ensuring your healthiness. You can sell surplus produce at the local farmer's market to generate a passive source of income.

15. Complete Design Layout

This layout has everything an off-gridder could need: a nice three-storied home, a sign by your home to advertise that you sell produce, a storage shed for tools, a relaxation corner, a hundred feet farm, a compost area, laid hedges, deep beds for small space gardening, a mini-orchard for fruits, a sturdy fence around your garden, a house for ducks or rabbits, a pond for the ducks and fishes, a beehive for your family, and last, a grazing area for pigs and other livestock. They thought of everything when they made this design!

16. The 2-Acre Homestead Layout

The two-acre homestead contains everything that you need in as little land as possible. For most of you, two acres would be more than enough to sustain an off-grid lifestyle; while for others, it might be too small. No worries, we're discussing all the possibilities here. For reference, two acres is roughly 90,000 square feet. You can check out the detailed infographic on the website. You can adapt a lot of plans for your land and customize the individual elements to suit your needs. This design is very thorough in explaining where to place each element.

17. The Hamilton Permaculture Layout

This layout has something that we have not considered before: a worm farm. Worm farms will give you great compost, which in turn will be advantageous to your gardens. There are bees, chicken tractors, aquaculture, and guilds in this design as well. The barebones of the design start off with a house, behind which is a sustainable backyard, next to which there's an adobe, a stacking, a compost pile, and a bed for gardening. The design has a luxurious feel to it, ergo the presence of "Hamilton" in the name.

18. Larger Farm Layout

This is perfect for those of you who have more than a few acres of land on their hands because it not only includes all the common farm buildings, but it also has a functional silo too. In this design, there is an orchard with a beehive and fruit trees, a pigsty, a bunker for the silo, the silo itself, enclosures for animals, a greenhouse, vegetable gardens, permanent pasture, a fallow, a meadow, an area for crops, and ornamental trees.

19. Urban Homestead Layout

This layout is a clever idea for those of you considering an urban design. There are functional areas and rest and relaxation areas on this layout.

20. Old School Homestead Layout

The old school homestead is reminiscent of the Wild West, complete with a ranch and all, but you must have a few acres lying around to pursue this layout.

21. All-Inclusive Small Homestead Layout

Perfect for multiple meat sources, this design doesn't rely on acreage as much as it focuses on functionality. There's also room for cold frame greenhouses. This layout utilizes all the space in a compact manner.

22. Family Food Garden Layout

Ideal for your family, this food garden layout is efficient because it utilizes all the space for growing a ton of food. If you have more than ample space available, it's probably best to keep the animals away from the house to avoid the smell and the sounds.

23. Family of Four Mini-Farm Layout

Most suitable for you if you have a small family, this design caters to a familial lifestyle, considering every aspect of homesteading that your family could require.

24. Modern Homestead Layout

This layout has an amazing planting space where you can plant edibles, fruits, and vegetables. Another huge plus is the modernity of the design, which will appeal to the esthetician in you.

25. Real-Life Layout

The real-life layout helps you in living a practical life, not that all the others don't. It's just that, realistically, you will have an easier time adapting to this design than all the other ones. You can easily manage all other aspects of your life without worrying about making too drastic of a shift, as other designs require you to do.

Now that we have covered every (well, almost every) layout for your off-grid home, let's move on to choosing your style of living structure.

Chapter 6: Choosing Your Style of Living Structure

Whether you want to reduce your carbon footprint or are simply interested in moving off-grid to live your life in secluded solace, you will need a house plan. A house plan differs from the layout because it focuses on the interior of the house. There are a lot of considerations to be made when living in your off-grid house, which we will discuss now.

Eight Features of Off-Grid Homes

The first thing you must remember is that off-grid homes must be efficient in terms of energy and water. There are many ways to go about it and we will discuss it in detail. It's all about the design of your home. Consider these eight things when choosing your design and style of living structure.

1. Insulation

The off-grid home should have very thick insulation, or at the very least efficient, if not thick insulation. It should be better than an average traditional house's insulation so you can conserve the sun's heat in the winters and keep your house cool in the summers.

2. Eaves

Eaves help you with heating your house when it's cold and with keeping it cool when warm. Practically the same function as insulation, but with a different method. Overhanging your roof so the winter sun—at its low angle—can get through your windows while the high angle sun's rays will not only reduce the energy consumption in terms of cooling and heating but also ensures that you get plenty of sun for your Vitamin D. The eaves shall be applied to the house's south side for those of you who happen to be living in the Northern Hemisphere.

3. Solar Light Tubes

Again, this is one of the best ways to save costs while remaining efficient. You can harness the power of the sun during the daytime to light up your home. Sure, the solar tube system will ensure the lighting for the daytime, but they're not intended for nighttime lighting. With that being said, it doesn't mean that they're not advantageous to have; they are, after all, saving you a ton of money in terms of power consumption, even if it is just for the daytime.

4. Cisterns

You need something to collect the water with, so why not cisterns? This way, you'll have water on your hands whenever it rains. We're not suggesting that you use that for drinking; but should you decide on using it as drinking water, filter and boil it first to purify it. Place plenty of cisterns all around the house to increase the surface area for rainwater storage.

5. Water Heater

You can either choose solar water heaters or tankless hot water heaters. The tankless ones only turn on when you need them to and promptly turn off when you're done consuming water for bathing or cooking. To step it up a notch further, you can choose solar water heaters for heating your water in the daytime. The only downside to this is that you won't get to use hot water at night, but that's okay, isn't it? You can take your baths in the morning ... no big deal.

6. Stove

A wood stove is a must-have in the arsenal of an off-gridder. It shall serve as the centerpiece of your house. Wood is probably one of the most effective ways to heat up your house, especially in those long, chilly winter months. When choosing a stove or, even better, building one, you must make sure that it's made of stone and is properly insulated to give you the most efficiency for burning wood. This is especially beneficial when, such as in winters, you have a short or limited supply of wood.

7. Ventilator

Not the kind that hospitals have. This ventilator is supposed to function as a heat recovery mechanism by capturing residual heat before it gets a chance to leave your home.

8. Solar Panel

If you install solar panels on the south-facing side of the roof, you'll get yourself a ton of solar power at your disposal. Remember, we're aiming to maximize our renewable power sources as opposed to other forms of sources, such as diesel. By placing the panels on the south side (no, not Chicago), you can get the most out of the sun in all kinds of weather.

Off-Grid House Plans

Whether you're looking to get started, or have already started with a modest budget on your hands, these following plans will give you a direction and an idea of how much budget you need. Let's get right into it.

1. Prefabricated ARK Shelter

The ARK shelter is perfect for those of you on a budget, as it comes at a relatively low price of $50,000, which includes the transportation and installation. Let's be honest with each other for a second. As an aspiring off-gridder, you probably have no idea on how to get started building your own home. That's okay. We don't need to reinvent the wheel when we have prefabricated options at hand. There are multiple styles that ARK offers. The most viable one being a cozy shelter designed for two people along with a cabin that's the length of a shipping container. This particular design also has a living room in the center, with just about enough space to fit a small dining table in there. Other key features of this home include a kitchen, a bathroom, and a bedroom. There's a huge panel window on one side, from which you can admire the view of nature.

Since it's a prefabricated setup, it comes with its own installed solar panels and can also utilize wind power should it be needed. There are cisterns on the roof to help you collect rainwater. There are different options for your sanitation, including septic tanks, waste chemical treating, and so on.

2. Minim House

The tiny house movement inspired the Minim House. You can use it on the grid as well to wade the waters before diving in. The house is small, with only 260 square feet, but there's room enough for practically everything. There aren't any walls inside the house, so the whole place gives off a studio apartment vibe. The bed can be pulled out of the wall and pushed back into it, thus giving you ample space to move things about. The kitchen layout of the house is roomy as compared to the rest of the house.

A series of solar panels on the roof provide a power of over 900 watts in terms of collecting energy. There is also a set of batteries built-in the structure, giving you the freedom to be totally off-grid. The batteries alone aren't enough, so you will need additional power sources including solar power, as discussed, and generators. Unlike the ARK home, there are no cisterns in the Minim house, so there's no way to collect rainwater, but there's a 40-gallon tank that's tucked away somewhere below the couch. There is also a water filtration system attached to that tank. There are big glass windows that will be the primary source of heat in the daytime. The doors serve the same purpose. The one downside—other than the lack of cisterns—is there is no wood stove to heat the house centrally.

3. Off-Grid Shelters

Suppose you are still a little apprehensive about the process of building your own homestead while at the same time not in favor of pre-fabricated homes, no need to worry; there is a third option for you in the form of off-grid shelters. It's recommended that you consult with an architect or a building firm that specializes in this exceedingly kind of shelter. You have a book of plans at your disposal, which the firm shall provide. This will inspire you to be creative with the interior design of your shelter and not just the interior décor. You can also be quite involved in the exterior design, tailoring each part according to your aesthetic and choice.

4. Tiny-A-Frame

This is your chance to have a trial run at the off-grid lifestyle without investing too much of your money, as the Tiny-A-Frame costs less than $1000, particularly when you're using recycled material. The cabin, as the name infers, is tiny—uncomfortably so—and covers just about 80 square feet. There's only enough space for a bed in there. There aren't any kitchen or bathroom units, so you must depend upon exterior resources for both. There are solar panels at the top of the frame house that provide little energy, but seeing as how this is such a tiny home, don't be worried about that. Like the previous house, the heat is provided by the windows and the doors. You won't be able to occupy this space in the winters, as the lack of a central heating system will make it difficult to bear the harsh weather.

5. Penobscot Cabin

This is the best plan for you if you are feeling confident about building your own house. Not only is this plan easy to build, but there is also so much potential in the way of adding new modules and areas to it once you have started off your homestead and have been running it for a few months. By default, this cabin does not come with cisterns or solar panels, but you can add them later. Other than that, there are no large windows either. Since there's going to be wood used in constructing this cabin, you will have natural insulation, which will come in handy in the winters. If you're skilled in terms of architecture and building, you can add on more modules once you've become comfortable building upon your cabin.

What About Underground Houses?

If you admired the books and movies featuring Hobbits, you will absolutely love the idea of having your house in a hole in the ground or under the hill. If you think that the homes would be dingy, dark, and smelly, you're mistaken. There are as many types of earth-sheltered homes as there are of the aboveground homesteads. Let's look at three of them for starters.

1. Earth-Covered

These types of houses have a living roof above them. They're going to look very much like normal homes, but without any roof. If you've got an earth-covered home under the hill, it shall serve as a cave of sorts, covered from three sides by the hill, and the fourth side shall have a door and windows on it.

While it is not entirely underground, this home is a primer in getting you started in living the underground lifestyle. It will ease you in, so to speak.

2. Earth-Berm

The earth-berm houses enjoy the popularity factor. Their styles vary from house to house. Some houses might have half-covered walls, while others might have two to three sides covered by the earth. Mostly, these houses are set in hillsides like Hobbit-holes and look quite regular when seen from the front. If you're the kind of person who doesn't like the aesthetic of subterranean houses, this might be your best bet.

3. Subterranean

This one is completely underground and has none of its walls made of brick and mortar. All four sides and the roof are covered by earth. Most of these houses are built into hills, just like earth-berm ones. They're an awesome place to stay, but again, one style might not suit all. Note that most of these subterranean houses offer you the luxury of traditional homes within the ground. From the inside, minus the windows, you will notice nothing that will suggest that you're living underground. Another advantage is natural insulation. You can save on heating costs by adding deep insulation to it. There's also the option of adding windows on the side with the door on it.

Advantages of Earth-Integrated Houses

1. Save on Power Bills

Depending on where you live and the weather there, you know how exorbitant air conditioning and heating bills can get. When you have an earth-sheltered or earth-integrated home, you will end up saving a fortune on your bills, since it will be buffered from all sorts of harsh weather from all sides. There's going to be coolness in the environment in the summers and hotness to it in the winters. It's a win-win as they say, right?

2. Natural Protection

Having a subterranean house will give you natural protection against most calamities, such as tornadoes and hurricanes. Since you're practically underground, there's little chance you will be affected by those calamities. That alone is the best and biggest factor you should consider when opting for an underground house.

3. Little to no Insurance Required

You'll find that insurance companies won't be taking a ton of money from you, as they know that there's going to be a lot of costs saved up in the form of natural protection. You're basically making their job easy for them. Traditional homes have a dozen natural threats posed to them, which subterranean houses do not.

4. The Cheaper Option

These houses don't need a lot in the way of architectural design and up-front costs, making them way cheaper than any aboveground homes. Also, since you're not using a ton of resources in the building process of the house, you're saving money on that front.

5. Cheaper Maintenance

Since this house has no exposed walls, there's no maintenance required. No repainting, no gutter cleaning, no refinishing, and no major renovations. They're not only easy to build, but they're also dirt cheap (see what I did there?) to maintain. You can save that money for designing the interior of the house.

6. They're Quite Unique

How many people do you know with underground houses, fictional characters excluded? You can spark conversations by telling people that you don't live in a traditional home and how it's better than their homes. There are online communities where people come together and contribute to the knowledge-bank of underground living tips and resources.

There are two things you must remember when choosing an underground house. First, there's going to be an extensive amount of effort that you will put in the construction and finalization of the house. Second, a lot of contractors and builders do not have experience in building these homes.

Chapter 7: Water and Sewage Options

Water and sewer are essential to a comfortable home, whether you live off or on-grid. There are more water options than there are for sewage, but all will be discussed.

Water Options

Here's a mindboggling statistic: the average person consumes anywhere between 80-100 gallons of water per day. Besides electricity and food, your main concern should be your water consumption. Realize that as someone approaching an off-grid lifestyle, you must remember the scarcity and the renewability of your resources. You're not going to have as much of anything as you did back when you were living in the city. The learning curve is a little too steep with this nugget of wisdom. Another thing to keep into consideration is an emergency. What will you do if an emergency occurs?

Let's discuss six options for your water supply and storage, so we're prepped for anything that nature throws our way.

1. City Water

Most homes built within the city will need to be connected to the city water system. You will have almost no choice in paying for the connection due to limited acreage and city requirements. Many HOAs will decide whether you must be connected to the city if you are within city limits. Mortgage companies are also particular on this point. Mortgage lenders want predictability when it comes to water. Unfortunately, it can come with downsides, such as chemicals to purify the water, being expensive, and becoming contaminated on a large scale. As someone who wishes to live off-grid, chances are you will not choose a homestead within city limits.

2. Wells

Wells are expensive to dig. It takes somewhere around $10,000 to dig one well, but they're self-reliant and independent of all sorts of sources. You're getting your water from the earth. Not the city. Not through pipes. Not through anywhere else. Plain, pure, raw earth water will always at your disposal. It's an excellent option to consider if you have a lot of land and have consulted with a geologist who, in turn, has told you that there's a water source running under your ground. A huge plus of this source is that there's no need for plumbing, no need for cisterns, or storage. All the water you need can come from underground, you can use it as much as you like, and the rest will remain reserved on the wall. It's an almost endless supply for a small cost. Did you know that your city water might be contaminated? This is hypothetical, but if it was to be, you're getting free from that contaminated water by choosing the well—which is as pure as water can get.

3. A Hand Pump

A hand pump is handy to have at your homestead for those times when you can't rely on your well all the time - or if it is too much work. Most wells come with pumps of their own, so there's no need to worry about having to install one all by itself. There are electrical pumps available with electrical wells these days, which are recommended. But if you don't have those available at your property, choose hand pumps, as there's no satisfaction greater on your homestead than doing something with your own hands. Invest in one close by your homestead, so you don't have to go all the way to the well to get water. It's one of the most budget-friendly options available to you.

4. Store Your Own Water

This sounds quite simple and straightforward, doesn't it? The truth is a commonly overlooked solution, so we must bring it up as the sixth point. This is critically important if you have a water source on your property, such as a river, stream, or a freshwater lake (even wells). You need to store all that water somewhere, don't you? We're not talking cisterns right now. We're talking about water jugs, containers, and other forms of storage that you can keep at your home. This isn't a long-term solution, mind you. This is a just in case for when there's an emergency, and you're cut off from all sources of water and power. Most of the off-grid living entails you preparing contingencies for when things go wrong. It's not necessary that they will, but just in case they do, you're prepared.

5. Rainwater

If you are living in a place with a lot of rain, this should be your go-to option. Now we're talking cistern and storage tanks, purifiers, filtration systems, and all that jazz. If you're getting constant rain, then this can be your long-term solution for water, but if you only get rain in spells, consider other alternatives. Be warned, though, that if you use that rainwater for drinking and cooking purposes, you will have to make sure that it's safe first. You must install a purification system specifically for this purpose. Always use this form of the water source with an alternative. This can't be the only source you rely on.

6. Hauling Water

Consider the worst-case scenario (yet again): your property is on dry, parched land with no well, no rain, and no other source of water. What are you going to do? Do you remember the bit where we talked about preparing for the worst? This is the worst. What are your next steps? Here's how you get started. Buy a huge tanker and attach it to the back of your car. You're going to take a trip to town and fill your tanker at the water station. You're going to take it and come back to your property and fill it in reserve dug in the ground. This is to keep the water protected from sunlight and to keep it cool. This is a tedious manner of work, but it's worth the work if you're in a land with scarce water.

Last, as a bonus, we'll discuss condensers. Condensers make water appear out of thin air. Literally, if you're living in an area that gets a ton of humidity, get a condenser to turn all that humidity to water.

Sewage Options

Living off-grid requires distancing from city water and sewer. If you are hooked into city water, then you will be sending your sewage to a city treatment plant. It is not all bad to use city sewer. Once you flush, you don't have to think about it again. Going with off-grid options requires more cost upfront, but you save on monthly payments.

1. Septic

A septic system either requires a vault or a leech bed or can involve both. Septic tanks or vaults must be large enough to sustain your household. For five people, a three-vault system is recommended. The type of soil you have also dictated if you need a pump that will get the sewage into your vault. Septic systems can run you as much as $10,000 but are often closer to $5,000. The price is always based on vault size, septic pump, and in-ground plumbing, plus labor to dig a hole for the system.

Once you have septic established, you have extraordinarily little maintenance to worry about. One of the oldest tricks in the book is to dump yeast in a toilet once a month, and it will keep your vault or tank exceptionally clean. If you do not use something like RidX or yeast, you will need to have your vault emptied every few years. This can be costly.

2. Latrine

Whatever you want to call it—latrine, outhouse, poop house—it is a hole in the ground you dig and place a shack over to keep your butt warm and off the ground. Eventually, you need the outhouse pumped or to fill it in and move your hut a little distance. Going with this option means outdoor excursions—not ideal—in cooler climates.

There is nothing wrong with an outhouse if you don't mind going outside and getting it pumped out every so often. If there are no HOA or city limits laws to interfere, you can live off the grid this way.

3. Camp Toilet

RV toilets are another option. Unfortunately, this method requires you to empty the small holding tank quite frequently. You also must live near a campground or have a larger hole where you dump your wastewater.

4. Compositing Toilets

There are at least three composting toilets to choose from such as: Humanure, Clivus, and Commercial. The Humanure bucket is something you can create yourself or buy pre-made. The Humanure composting option has two stages, where you collect your waste in a pot or vault, like a latrine, but usually by using a bucket. A standard bucket with a toilet seat and wood box to hide it plus an enclosure with a small fan for a vent ensures you have comfort without the smell. You will need to cover material to help keep the smell down. Woodchips, sawdust, peat moss, dirt, or clay can absorb the liquids and reduce the smell. About two times a week, you will need to add the bucket contents to an outdoor compost pile. After you dump the material, it will need time to cure, and then you can use it for your garden. This option can cost you as little as $20 for the bucket.

Clivus composting toilets range in cost from $100 to $1,000. They are pre-made compositing toilets made for those who wish to live off-grid. It works in the same way as your homemade option, where you have a waste container with an absorption material; but you can build it into your home so you can empty the toilet less often by placing a wheelbarrow underneath.

Commercial composting toilets are more than $1,000. They are plastic and like the Clivus. You can get about 80 uses out of the toilet, dumping the liquid once a week. Typically, peat moss is the best system, with a urine separator attached. You need less peat moss because you separate the urine. A few commercial options include Nature's Head, Separett, and Sun-Mar.

5. Incinerating Toilet

If the idea of dealing with your waste is not favorable, consider an incinerating toilet. You do not need an expensive septic system or a way to empty your solid and liquid waste each week. An incinerating toilet, like EcoJohn and Incinolet, will turn the waste into ash. The downside is the energy it takes to incinerate your waste. There are gas and electric models. Both types will burn the waste until there is just a little ash left, which you need to bury. The burn cycle takes about an hour, but you can still use the toilet while it is burning waste. If you are using solar power, you will need to account for higher energy usage with an incinerating toilet. For gas models, you will need more propane than if you went with septic, composting, or outhouse sewage options.

Of the above options, a septic system is the most effective for living comfortably off the grid. Animals go wherever they wish in nature. They don't even bury their waste. I don't recommend this idea because floods or other issues can create problems. You are better off with some type of vault system, even if you compost your waste.

Chapter 8: Heat and Electricity Options

Planning for heating and electricity is a smart choice regardless of where you will build your homestead. You want to decide on the heating and electricity for your home to ensure you are designing it for optimal energy efficiency and comfort. You will have plenty of choices for heating, although some are more cost-effective than others. We will begin with heating options before talking about the two most common off-grid energy options.

Heating Sources

Commonly, electricity, gas, oil, and wood are used to heat homes; however, you do not have to stick with those choices. Instead, you can choose among numerous options to heat your environment and ensure your electricity, gas, or oil usage is kept at a minimum. Living off-grid is not only about creating a minimal impact situation but also about low-cost.

1. Electric Baseboard Heaters

Electric baseboard heaters rely on a source of energy, wind-powered or sun derived. As an independent system for each room, you will need to determine how many heaters per square foot are necessary to keep the room warm. Account for any sunlight that can improve the warmth of the house. Small rooms like bathrooms may need one heater, while larger bedrooms and living rooms require two or three heaters for an even distribution of warmth.

2. Central Heating and Air

A heating and air system, usually called HVAC, has a central furnace, which pushes warm or chilly air throughout. The air goes through metal vent shafts and is distributed into each room. It has one thermostat, which can have an energy-saving property, where you set the temperature for when you are there or away. These systems are costly and require more energy to run than some of the other options discussed in this section. You will need to provide yearly maintenance and change out the filter every three months.

3. Boiler

A boiler works with propane and water. A boiler works like an electric furnace pushing air throughout the house. The water will evaporate as it is heated to create warm air that will then go through pipes or metal vents to heat each room in a house. Boilers require yearly maintenance to ensure carbon monoxide poisoning will not occur. Like central heating and air systems, boilers are expensive to install.

4. Corn or Pellet Stoves

Corn stoves came to be during the depression when coal was too expensive to buy. Families burned corn and found they could heat their living space for a lower cost. Corn burning or multi-fuel pellet stoves provide 8,000 BTUs, which is like the wood pellet option available. Depending on where you live, you can find corn is more cost-effective than wood pellets.

5. Masonry Heaters

Masonry heaters are like wood stoves; however, instead of needing a fireplace, you build a brick stove, with chimney, and burn wood directly on the stone. Wood stoves can lead to a creosote fire, whereas masonry heaters reduce this risk. Masonry heaters use a renewable fuel, they can fit your home size to ensure you get enough heat, and they are easy to operate compared to a wood stove. Masonry heaters are also exempt from fire bans because they burn cleaner. There is little maintenance required, only the occasionally emptying of the ash and cleaning the chimney. The downside is that you need to be there to monitor the fire, so it is not usable while you are away. You also need to make sure the build is airtight to get the most out of the heat produced.

6. Fireplaces

Fireplaces or wood stoves use wood to create heat. You usually need to install a blower, so the heat radiates further into a room. They work great in the room they are in or if you have a two-story homestead with an open floorplan upstairs. You need to be home to be safe, and you are subject to fire bans. You will need to clean the fireplace of ash frequently based on how much you use the stove. The chimney will also need to be cleaned regularly.

7. Air to Water Radiant Heat

Radiant heat is not a new concept. Like the boiler, you use a system to warm the water. The difference is there is a radiator, typically, in each room that uses water to produce warm air.

8. Solar Window Boxes

Solar window boxes warm the air instead of vents, which is then pushed through the house like central heating and air system. The key to solar window boxes is to make sure the boxes have direct sunlight. You will need a translucent box that will warm in the sun. The warm air will enter the home via the window or other opening and heat a room. You would need one in each room. You should not depend on this as your only source of heat, depending on where you live.

9. Home Design for Heat

It might seem counterintuitive to have more windows in your design. You might think the frigid air will make it too cool and take away from the heat you want to keep inside. Depending on where the sun hits your home, having a wall of windows can ensure you have more heat coming in during the day. Like the window boxes for solar-generated heat, a wall of windows in direct sunlight will heat that room or multiple rooms on that side of the house. Rather than running heaters and using more energy, you can use the natural warmth of the sun.

10. Biomass Systems

Using your composting pile, you can also use the sun to generate heat. A composting heap in direct sunlight will become very warm. If you have a biodegradation setup with coils of pipe, you can feed the warm air generated by the compositing pile directly into your home. The biodegradation system ensures pure air reaches your home. The amount of compost you have or the number of piles you have determines how many rooms you can warm-up.

11. Using Natural Landscape

Some of the most interesting tiny homes, with off-grid setups, use the natural landscape around the house to generate warmth. For example, if you build into the side of a hill or mountain, where dirt and rocks basically cover one portion of your house, you can keep more heat inside. The design requires concrete or masonry work. One house even created a roof garden, allowing food and even moss to grow. The sun heats the material and, in turn, keeps the heat inside. With windows facing the sun, the home with its concrete walls and the covered roof kept the owners warm. You can also do this, depending on where you want to live. Concrete, adobe, or brick homes bake in the sun, which helps keep the house warmer on the inside. Darker stain or painted walls also help generate more heat in the sun, which can ensure the inside is warmer when the sun is up.

12. Oil Based Heaters

Radiators, oil lamps, and oil heaters are three ways you can generate heat by using oil instead of propane, wood, or the sun. To heat your home while you are away, you need an oil radiator. Oil lamps and heaters should be used only if you are home unless they are similar to baseboard heaters.

Electricity Options

There are two ways you can live off-grid and produce your own electricity. It does not mean you cannot live off-grid with just these two options for light and heat. Oil lamps are one way to have lights without electricity. Candles are another. The drawback is the harder it makes a living. When cooking without electricity, it means lighting a fire in a woodburning stove or making a campfire. You could have propane for cooking, and that would also allow you to use propane methods for lights and heating. However, let's assume you would rather use one of three energy power options.

1. Solar Power

Solar power is not as simple as installing panels on your roof. Solar power involves photovoltaic panels, an inverter, and batteries. The batteries store the energy you create with the photovoltaic solar panels. You also need plenty of sun or indirect sunlight. If you live in a place that does not generate enough sun energy, you may find you do not have enough electricity in your home to warm it, cook, and use lights.

While most of us want to be off-grid, and without a mortgage, you may decide a mortgage is necessary to get the funds to build your home. You cannot be totally off-grid with a mortgage. Banks are unwilling to lend to someone who uses solar power as the only electricity source. Keep that drawback in mind as you plan your home.

Before you can get a price for your solar power system, you need to know what components will use the energy source and what they draw. In electrician terms, you need to understand the load and run time of appliances, water heaters, lights, and other electrical components in your home.

Once you have the components you want to run and the run time, you can calculate the watt-hour. You will add all the watt-hours for every electrical device together. Most systems will lose about 30% in the generation and storage of energy. So, with the total watt-hour, plus the estimated loss of 30%, you can then calculate the load.

There are different solar panels, including mono and polycrystalline. Check the size, quality, and type to determine which option best fits in your budget and maintenance needs. Your battery selection is costly, and you want to ensure you have enough battery to store the energy, while also choosing quality. The batteries can run $400 per battery, so you want something that will last 8 to 10 years, rather than shelling out $400 every 2 to 4 years.

The next components are an inverter and charge controller. The charge controller sets the voltage to ensure no one appliance or circuit is becoming overloaded. The inverter will convert the solar energy collection into usable electricity. You have a direct current from the solar panel that needs to become an alternating current to run your appliances.

A lot goes into solar power, but if you have enough panels and calculate your usage correctly, you can have enough electricity to power your home and even have some leftover.

It is best to have a certified solar power expert help you with the calculations and equipment choices or to purchase a guide that goes into more depth about installing solar power.

2. Wind Power

Wind-powered homes are secondary options that work great in places with less sun and more wind. Some cities do not allow wind power. HOAs can also have issues against wind power due to the less than favorable appearance of the turbines. If the wind does not blow, you will not have electricity. You also have moving parts on a wind turbine, which requires more maintenance and costs than solar power, but the benefits can outweigh other options if you have plenty of wind.

Like solar, you need to know the load to correctly calculate the size of the turbine you need to power your home. Size will matter. A 400-watt wind turbine will handle a few appliances, but not an entire home. 900 to 10,000-watt turbines on a 100-foot tower can handle an entire house. You could have two or more turbines, but most often, going with one that does the entire house is more cost effective than having multiple turbines and towers.

3. Micro-Hydro Electricity

Micro-hydroelectricity is not as often discussed, but it is still a choice that works if you live near plenty of running water. Micro-hydro energy derives from a stream, river, or another consistent running water source. Energy comes from the water that flows from an elevated level to a lower level into a turbine. If one has a good water source, the energy can last for 24 hours, 7 days a week, and ensure you are never without power. You also require fewer batteries as part of the system because of the constant production of energy. Unfortunately, if you do not have a stream or river in your backyard, then you cannot use this option.

Each of the electricity options requires you to have a home plan designed for the conservation of energy. You want to make the most of your home plan, so you get direct sunlight and warmth from the sun. You also need to choose appliances with a high rating of energy efficiency to reduce the energy usage and therefore load on your energy system.

Chapter 9: Gardening for a Food Supply

Gardening for your food supply is a part of living off-grid. There are several benefits to growing your own vegetables and fruits. You have control over how they are grown, what natural pesticides you use, and the ability to keep your food GMO free. It is important to start any garden with non-GMO heirloom seeds. They do cost a little more, but the benefit is worthwhile. As we assess this topic, we will examine how to create a compost pile, keep your garden organic, and control pests. We will also discuss the best types of gardens for drainage and maintenance, along with some foods you might wish to grow.

Optimizing Your Garden Space

The most important part of your garden will be the planning stage. You not only need to determine the space you have for gardening, but you must also assess the sun, soil, and size. The more space you can allot to a garden, the more you can grow, but you can also find yourself where you are overusing the soil.

There are plants that cannot be near each other. They will compete and die instead of growing into luxurious fruits and vegetables. Tomatoes, squash, beans, and cucumbers are at least four plants that grow upwards and will need stakes or trellises to ensure they will grow healthy.

Mint overtakes a garden. As an herb, mint will grow quickly and vine along with the earth, overtaking any space you provide, so it requires constant care to ensure it does not cover other plants.

Strawberries also vine and can continue to grow throughout a row, expanding until you have too many. So, the following assessment will be necessary to successfully grow your plants:

1. Look at your land.

2. Where does the area get the most sun?

3. How much square footage can you devote to the garden without losing space for your home or other things you'd like?

4. Do you want to build a greenhouse?

The above four points help you optimize based on space and weather. If you live in a cold, snowy climate, an outdoor garden may produce for three months out of the year. This may not be enough to feed your family for an entire year. A greenhouse is a clever idea for small gardens that need to produce year-round because you have more climate control inside.

Another option is to have a winter/summer grow table, where you will cover your garden with plastic sheeting in winter and shade cloth in summer.

1. Garden Size

The size of your garden depends on several things, the first being the amount of land you possess. With just one acre of land for a garden, you can grow enough food for 30 people for an entire year.

You may not need a full acre if you are just trying to feed a family of four and have food leftover for sales. You can monetize your garden, which is discussed in a later chapter. For now, consider the acreage of your lot and how much food you need to grow to feed just your family.

The optimal garden size allows you to grow what you need without creating a feeling of scarcity. For example, an entire 30-foot row of lettuce might be too much for one family. Six plants can be enough for a weekly salad for five people.

2. Types of Plants

Grow what you know you and your family will eat. For example, if you do not cook with parsley, don't grow it. If you are carrot eaters, make sure you have enough plants to account for how many pounds of carrots you eat per year.

The types of plants you choose should be based on garden size, growing seasons, and your preferences. In mountain areas, tomatoes, blueberries, and beans grow well, while warmer climates support more diversity such as raspberries, strawberries, lettuce, cucumbers, and much more.

If you will plant directly in the ground, choose hardy plants that can withstand the frost. Also, research what each vegetable or fruit requires for sun, water, and plant food. Some plants need more or less sun than others. You want to make sure you keep like plants next to each other without creating a situation of competition.

Chives, dill, tomatoes, and peppers grow tall, while mint and basil are medium height. Lettuce is a shorter plant, as is parsley.

You do not want to grow a tall or medium plant near a shorter grower as it can shade the other plants too much and prevent growth.

The growth time should also determine how you plot out your garden based on types of plants. Carrots take several weeks to grow from seed, and even after 100 days, you may not have a very large carrot bunch. Whereas mint grows within a few weeks and will continue to expand unless you trim it every week.

Regarding types of plants, it may be in your best interest to get a book or guide that discusses each vegetable and fruit best suited for the growing conditions before you put it in your garden.

Types of Garden Beds

Raise beds, inground, and container gardens are three types of gardens you can create. Choosing the correct style will ensure that you have enough plants healthily growing to feed your family. There are seven types of growing beds, plus container gardens to discuss.

1. Raised Beds

A raised bed garden is where you build a box, place your soil inside, and then put your seeds in the planter. It can also be a mound of soil that is higher than the dirt around it. The idea of a raised bed garden is to ensure proper irrigation and soil medium for the best growth potential. Usually, you have one raised bed per plant type.

A raised garden bed allows a deep growing area where the plants roots grow down and out. Another advantage has the beds at eye level, which helps you take care of the plants without hurting your back. A higher bed can also help you see the pests better. Native soil can be contaminated or improper for the types of food you want to grow. Raised beds to ensure you have the correct medium.

Raised beds are a permanent garden, where the soil is exposed to heat and cold. If you build your beds you'll want to have thin walls as well as an irrigation system to ensure proper moisture because raised beds can become dry more quickly. You also need pathways between the beds to help your garden.

When choosing raised beds, you want to decide the ideal height based on the types of plants you are growing. A height of 12 to 18 inches is typically ideal for things like carrots that need at least a foot of underground growth. You also need to determine the width and length based on the number of plants you need to grow in the bed. Rectangles are the typical design; however, you can make T's, ovals, or circles depending on your landscape and other garden beds.

Some plants need 12 square inches per plant, which limits how many you can put in a raised bed rather than a mound raised bed style.

The wood you introduce into your garden for raised container beds should be untreated. Treated wood can have chemicals that will affect your plants. You want healthy soil, so going with natural untreated wood you stain on the outside is better. A raw treatment like linseed oil is even better than the stain. You want something on the outside that will help prevent wear from the weather. Overtime, sun on untreated lumber can cause it to become compromised. Typically, if you just get sun only on the outside edge, it should not interfere with plant growth. There are some studies that indicate paint will trap moisture in the wood that will make it rot overtime.

People have used a variety of products like block, composite wood, and railroad ties to make raised container beds. You are better off avoiding all these, including galvanized metal. They all have products that could create chemical toxicity even though current research is not detailed enough to show such a thing has occurred. Tires are also something to avoid despite how handy they might seem. Rubber is not an excellent product to contain anything you will eat.

Many people also line the raised container beds with plastic, which can cause health issues. You need to purchase food-grade polyethylene plastic designed for gardens to ensure there are no issues with plastic toxicity. Plastic can be helpful for a barrier between the bed and soil; however, drainage must be a consideration.

Mound raised beds to eliminate wood issues; however, most people believe in using a weed barrier to keep weeds from interfering with food growth. You do need to get a food-grade barrier.

2. Clipping Beds

Within the raised bed category, you have clipping and plucking beds, which house two types of plants. You might integrate some of your plants, so you would want a clipping bed filled with plants you will need to clip, such as chives and mustard greens.

3. Plucking Beds

Plucking beds are those where you remove the top leaves or flowers from the edible plant to keep it contained, but also to take what you need to use for recipes or to eat. Plucking beds usually have faster-growing plants like broccoli, kale, zucchini, cucumbers, and fruit plants.

4. Narrow Beds

Narrow beds are better for carrots, beans, tomatoes, peas, radishes, and other root vegetables. On either side of the mound, you bury the plant to give them 12 inches space. The root grows the food you will eat in the soil so the plant will have short tops.

5. Broad Beds

Broad beds are better for larger vegetable and fruit plants like pumpkin, sweet corn, cabbage, and cauliflower.

6. Herb Spiral

An herb spiral is a vertical style bed, where you use both the grown and vertical space to create an herb bed. The circular creation ensures plants have more sun exposure than others. You will also get different wind and temperature exposure for the plants. You want to have the correct soil for the plants you grow, and due to the spiral, there is water drainage from top to bottom. So, plants that require less water should be on the top.

7. Vertical Planting

Vertical planting beyond an herb bed can include using trellises, pergolas, and fences. Some people even use the side of buildings to grow their species of plants. Peas, beans, cucumbers, grapes, and kiwifruit are commonly grown in vertical garden beds. You start with a mount and then place fencing or a trellis around the plant so it can vine up rather than out. You get more plants with vertical growth because you can fit a plant every 12 inches rather than worrying about vining plants that need ground space.

Container Gardening

Raised beds and mound gardening that go directly in the ground are the best options, but talking about container gardening is important for those with limited space or seasonal needs. For example, if your landscape does not lend to gardening because you live in the mountains with too much rock and sloping ground, containers might be the better choice. In colder climates, you might want the option of moving the containers indoors during the winter season.

Like raised beds you construct, you want to be careful of what you choose for your containers. Buy containers like ceramic pots from the gardening section at your local store.

Also, realize you are limited in the space you have for containers. You might get one plant, such as one blueberry bush in a 100-gallon container. Herbs can handle smaller pots. You can have six plants in ten inches of planter without having problems.

The depth of the pots used also must account for the root system. For carrots, a deeper container is necessary. Strawberries are fine in a long shallow container.

Hydro-Gardening

The most famous hydro garden is the Aero Garden. It is a fully contained system that uses water, plant food, and ten hours of light to help you grow tomatoes, peppers, lettuce, and herbs. They also have a strawberry Aero Garden. While expensive to get the container, the plants are affordable, organic, and the yield is high. But it is unnecessary to purchase the small gardens to have a hydro or water-based garden.

You can create your own hydro garden using large containers filled with water and direct sunlight. When building the container, it needs to be watertight. Your other option is to do an inground hydro pool. If you live in a climate where you need not worry about winter and frost, you could dig a large hole that would make a small lake or water trough. It needs to be made to prevent water from leaking into the surrounding ground. It need not be deep. Only about six inches to a foot of water is necessary for most plants. A pump is necessary to circulate the water every few hours. The garden requires at least 10 hours of sunlight or plant lights. Pack the starter seeds in a growing medium and let nature take its course.

For smaller water gardens, you can use containers and a large fish tank. A pump will circulate the water, while you can have water pumped into a variety of containers from the holding tank. The fish tank would be the holding tank. On a cycle, the water circulates goes through a filter and goes back into the holding tank. With individual containers and a holding tank, you can the plants individually rather than having one giant pond filled with a variety of roots and plants.

Water Catchment Systems

Whether you choose a hydro-garden or a raised bed design, you still need water to help you grow your food. Living off-grid usually means a well and septic for your needs. A well might produce as much water as you need without ever drying up. However, sometimes a well needs time to refill, which could mean not enough water for showers, washing dishes, and other basic needs, let alone watering your garden.

The amount of rain you receive in your area per year also determines how moist the soil stays. Water catchment is one way to ensure the garden has enough moisture, during dry seasons especially. A 500 sq ft roof can capture about 300 gallons of water during an inch of rainfall. For a small roof, that is a good start toward ensuring your garden is watered thoroughly when rain is lacking. There are two ways to make water catchment work.

1. Let the rain roll off the roof, through gutters and let it run downhill into your garden.

2. Use gutters and a rain-collecting device, such as a barrel, that you can then transport the water to the garden when you need it.

There are other water catchment options. Just sitting barrels out in the field can ensure you collect rainwater as it falls. Although many find a roof, even a shed roof is better for catching more water. With hydro systems, the pond or holding tank you have can be the catchment device making it easier to keep your garden growing.

Irrigation

With catchment systems that include a barrel for ground or raised bed gardens, you will want a pump that will help you distribute the water throughout the garden. You can use an elaborate irrigation system, with pipes running in the ground and sprinkler heads that pump the water onto the leaves. There are also inground choices that do not have sprinklers and simply feed the roots.

Irrigation has been around for centuries, so living off-grid with an old-fashioned pipe system is possible and quite reliable.

Drip irrigation requires 10 to 30 psi (water pressure), but even 8 psi can be enough. You do not want too much pressure because the leaves and plants could become damaged.

You may also consider using a river or stream nearby to help you get water to the garden via a pump. If you live in an area without rain or where water catchment is not as great, but you have a river or stream, then pumping water from the stream, through pipes, and into your garden is another way to ensure you are getting the proper amount of water.

Pest Control

Moving on to pest protection, gardens are always a "hotbed" of desire for various pests. Deer, boars, bears, and smaller pests like insects and groundhogs can be very detrimental to your garden. Living off-grid, you probably want to keep your food organic, given it is healthier, which is why it pays to ensure proper pest control is available.

1. Solar Fencing

Solar fencing is for the larger rodents and pests that might try to eat your garden. Any fence you put up around your garden should be buried at least a foot into the ground. The deeper you go, the easier it is to keep rodents out. Rabbits, foxes, and raccoons won't dig deeper than 12 inches. Groundhogs and other underground dwellers are a different matter. They can burrow under the fence and attack from the bottom. Solar fences are also used to keep deer and other grazing animals out of the garden.

Solar energy powers the fence to give a small electrical jolt to the larger pests. You want a tightly woven fence, such as chicken wire, to help keep out the medium-sized pests. By using electrical wire and a solar panel, you can electrify the fence.

Deer and elk can jump fences, and you may not want to make it too high. It sounds cruel, but baiting the wire with things deer like to eat helps them realize they do not want to touch the fence. This can backfire. Deer and elk are such great jumpers they may not have to even touch the fence to get over it if it isn't built high enough.

2. Natural Insecticides

Keeping small insects out of your garden can sometimes be harder than the large pests. You don't want to use pesticides that will harm you and your family, so learning about the 8 natural insecticides will help you come up with a concoction that works for the pests you will encounter.

1. **Soapy Water** – 5 tablespoons of dish soap in 4 cups of water kills aphids and mites.

2. **Neem Oil** – Readily available as an essential oil—neem is derived from trees that grow in India. It is an anti-fungicide, kills scale, mites, and aphids, and other insects.

3. **Pyrethrum Spray** – Created from chrysanthemum flowers, the powder should be mixed with water to help stop flying insects.

4. **Garlic** – A vegetable not only helpful for cooking, but also stops bugs. Be careful with growing garlic with some of your food plants though since they can compete.

5. **Beer** – And you thought beer was just for drinking. Beer is made with yeast, barley, and other grains, which you might think attracts bugs. However, due to the alcohol and other ingredients, insects are more attracted to a saucer of beer than they are your plant root, leaves, fruit, and vegetables.

6. **Pepper** – Some people use pepper spray using a red pepper, dish soap, and water. Any pepper, including paprika, helps stop spider mites and other aphids. Take 2 tablespoons of pepper, 6 drops of dish soap, and a gallon of water, mix it, and put it in a spray bottle.

7. **Herbal Sprays** – Thyme, sage, rosemary, rue, lavender, or mint crushed and soaked in water can make an herbal insecticide. Even putting these herbs in your garden and growing them for recipes can help your pest control efforts.

8. **Nicotine** – Using 1 cup of dried tobacco leaves and a gallon of water ensures you can keep leaf-chewing bugs off your plants.

The nicety about using these different methods is that they are safe. With a thorough washing of your plants, even dish soap is less harmful than the commercial pesticides sold for gardening.

Composting

How will you feed your garden? Soil may have enough nutrients for one year of growth, but after that, you will need to add some healthy food to your garden. Composting is one way to keep your garden growing long after the original soil has lost its nutrients.

An area of your land should be set aside for a compost pile or three. Three is best—one you use, another ready for the next year, and a third you add to throughout the year.

The benefits of composting include:

1. Cost-effective due to you and your animals contributing to the pile.

2. It is a natural way to ensure your garden soil has the appropriate nutrients.

3. Your compost ensures the healthy growth of plants.

4. Composting reduces the pressure on landfills.

5. Composting also adds to water conservation because the soil can retain more water.

6. Composted materials have natural fertilizer.

Compost Ingredients

1. Straw

2. Leaves

3. Sawdust

4. Newspaper

5. Wood chips

6. Grass clippings

7. Vegetable peels

8. Tea leaves

9. Eggshells

10. Coffee grounds

11. Weeds

12. Fruit

13. Wood ashes

14. Garden residue

15. Waste

The size of your garden determines if buying or making a compost bin is better than having a large pile of ingredients taking up a 100 sq ft space. The key to any compost pile is to ensure it has plenty of oxygen and moisture to work properly. The matter needs to be broken down, so it works as a compost. Any compost you use should have the moisture level of a wrung-out sponge. It is better to go for too dry than too wet. Too wet can lead to issues of mold.

Types of Bins

1. **Tierra Garden** - Created from recycled plastic, this 80-gallon bin helps you compost kitchen and garden waste. It is prebuilt to help you compost, while also keeping it tidy and critter-free.

2. **Wooden Box** - Wooden building boxes can be another way to create piles and keep them tidy. Old pallets, discarded building materials, and logs have been used to make composting bins. The one thing you want to take care of is to use untreated and unpainted wood. You don't want to add chemicals to your compost through the wood you use to build the bin. With a compost pile, the bottom will rot before the top, so shallow bins you rotate the use of will ensure your compost remains usable.

3. **Countertop Composter** – If you have a smaller garden, a countertop composter might work better. These bins are expensive because they are made with recycled plastic and contain compostable bags to keep the process simple.

The size of the bin determines how long it will take for your compost pile to be ready to use. The process can take as little as 3 months in small pins and up to 2 years if you just keep piling things on top of the original pile.

Aeration is necessary for the oxidation of compost materials. You do not have to use a bin, especially if you want to turn the pile so oxidation occurs and helps the microorganisms break down the material. Aerating compost every few days is the best routine.

Watering your pile depends on how humid and warm your climate is. A dryer climate will require more watering when in the direct sun versus a moist climate that may never allow the pile to dry. A soggy pile will rot and not go through the proper aeration process. Instead, it will mold, and if you use the compost, it will cause mold to grow on your plants. Vegetable and fruit mold differ from the mold that grows from being too wet. If you have aged fruit and vegetables with mold, those mold cells will help with the decomposition of your compost materials.

Uric acid or urea is helpful in compost piles. Get into a routine of dumping your waste from a composting toilet in the morning to allow it to move from the top of the pile and soak into the material.

Chapter 10: Raising Livestock for Food

Don't read if you are a vegetarian! Only kidding... You are reading this chapter because you want to know how to raise different livestock for food on your off-grid land. Raising livestock can be an expensive concept because it requires having enough space, includes slaughtering fees, veterinary care, maintenance, guarding, and raising enough to feed your family. If you are a beginner homesteader, start with chickens. They are low cost compared to cattle, goats, and pigs. Throughout this section, you will gain knowledge on raising livestock and slaughtering, particularly the crucial factors of local ordinances and costs for using controlled slaughterhouses.

Raising Chickens

Over 500 breeds of chickens exist; however, there are perhaps a dozen you will be interested in adding to your homestead. Ameraucana chickens are some of the best egg-layers you can ever have, plus the give you blue-shelled eggs.

Chicken Breeds

1. **Ameraucana Chickens:** A medium-sized bird that produces large blue eggs. Sometimes Ameraucana can lay more than one egg in a day and typically provide 250 eggs per year. Ameraucana have dark feathers, often with some silver and blue.

2. **Australorp:** A black chicken with a large comb, the Australorp was brought to the US from England, although they are an Australian bird. The Australorp produces at least 250 brown eggs per year and is even known to lay 364 eggs without artificial lighting.

3. **Bielefelder:** This is a dual-purpose chicken because it is a great egg-layer, providing at least 230 brown eggs per year, but is also a 10 to 12-pound fowl. Once it has reached the end of its egg-laying years, it can become food on your table to make room for new, younger chickens.

4. **Orpington:** Another English breed, it was breed by William Cook for eggs and meat. Orpington chickens can be orange-red in feather color, but also black, white, buff, jubilee, and spangled. Orpingtons are known for laying at least 200 brown eggs per year. They are also the leading chicken for those who want a pet versus a meal.

5. **Plymouth Rock:** A mostly black chicken with spangled white plumage; the Plymouth Rock is an all-American bird. It was created for its eggs and meat, plus its ability to adapt to hardy conditions. The Plymouth Rock is a brooder, with docile behavior. The Plymouth Rock also produces brown eggs.

6. **Rhode Island Red:** Another large bird, the red-feathered species, offers five to seven brown eggs per week and then becomes a great fowl for the dinner table.

7. **Cinnamon Queen**: The cinnamon queen is known for its brown eggs, but they can also produce white eggs since they are a cross between the Rhode Island red breed and white breed. This is a breed to have if you want egg laying to start quickly. Pullets lay 250 to 300 eggs per year, and typically start a week or two before other pullets born simultaneously.

8. **Barbezieux:** A French breed, the Barbezieux weighs 9 to 12 pounds in adulthood. While not breed specifically for egg laying, this breed does provide a significant amount of white eggs. However, it is the firm and distinctive flavor of the meat that farmers appreciate.

9. **Cornish Chicken**: Also known as Cornish Game Hens, you hear about these birds being a small, singular meal for people. The hens are usually harvested at one pound, considering this is when the meat is most tender.

10. **New Hampshire Chicken**: A red-feathered bird, the New Hampshire chicken can be dual purpose. It's a happy free-range bird that can produce a significant amount of eggs before you add it to a table meal.

Remember; there are plenty of breeds to add to your farm. The above ten were chosen based on knowledge of raising chickens and personal preferences.

Chicken Life Cycle

All chickens are hatched from a fertilized egg. You can buy fertilized eggs and help your chicks hatch, or you can purchase chicks to begin your homestead. An egg hatches after about 21 days. When a chicken hatches, it will have a wet "down" feather, which dries quickly and turns them into a fluffy adorable chick.

Chicks need to be in a brooder. A brooder is an indoor space with a heat lamp. In the first couple of weeks, clean the brooder each day and supply a baby with chicken feed. After five days, a chicken will show some real feathers, and by day 12, you will see a significant breed indication. By 18 days, the chicken will have most of its regular feathers with very few downy ones left. After a month, more breed characteristics appear.

Young chickens that have reached egg-laying status are called pullets. At 18 weeks, most pullets are ready to lay their first eggs. Healthy well cared for chickens will lay eggs for three to five years. The number of nutrients, warmth, and care you provide your chickens will determine how frequently they lay eggs. Their breed also indicates if you will see five to seven per week or a cycle of one egg every other day.

Chickens molt annually to remove old feathers and grow new ones, which can cause a slowdown in their egg production during the molting process.

Unless you intend on allowing a rooster to fertilize some eggs for new chickens, you probably want to buy already hatched chicks to avoid getting a rooster in your pen.

Many cities and even counties have laws against roosters. For example, one county states that a person can raise up to six chickens for egg laying or meat but cannot have a rooster. If you are homesteading and have a commercial license for livestock, the rules can differ. You may be able to have a rooster on a "farm," despite county or city codes. Research this aspect based on your city, county, and state.

Chicken Coops

Whether or not you will raise free-range chickens, you still need to have a chicken coop. A coop is a place for chickens to brood, ensuring you get the egg production you desire. Some chickens will not care if there is a nice pile of bedding to roost in and then sit on their eggs. These chickens tend to drop an egg wherever they are at the moment. Others are great brooders, so they will dig into the dirt or pile bedding up and then lay their egg.

A coop is also necessary for the protection in the dark hours. Coyotes and wolves are just two predators that enjoy a meal of fresh chicken. By providing an indoor space, you keep your birds safe and less stressed.

The coop needs to have a 1 sq ft space per chicken. It should have a vent, a ramp to get inside, and a larger door to help you access the floor for cleaning. A window and even a lamp are a clever idea to help keep your birds happy, in the sun, and warm.

Any coop you build or buy should be airtight. Think of it as a mini home where your chickens can winter warmly without a heat lamp. The cost of creating a coop can be as little as $100.00 when you use recycled materials, nails, and tools. Buying a coop can range from $100 up to $5,000. The larger and more elaborate the enclosure, the more it will cost. Fencing

Even free-range chickens should be contained within your homestead land. The idea of free-range is that the chicken can roam and eat the grass and seeds that naturally occur on your land, rather than being cooped up in a small building or small fenced paddock. Free-range chickens can mingle with other livestock.

Any fencing you have should be buried at least a foot deep. It should be "chicken wire," which is about a centimeter in size per square hole. This prevents smaller pests like raccoons from getting in and stealing eggs or harming your chickens.

If you have a lot of pests in the area, it can be a clever idea to have a wire roof over the fenced area. There are rolling fences that let your chickens graze on your open land without being vulnerable to predators.

Food and Water

There should always be water available for your chickens. There are a variety of water containers from automatic waters to a simple bowl on the floor. A wall container is better because the water remains clean versus a bowl on the floor that could be used for bathing, plus excrement. Some chickens use bedding to bury the bowl of water. However, if you live in a cold climate, you may need one of the floor bowls with a heater inside. The heater will keep the water from freezing during the cold months.

As for food, you will need to begin with chick feed if you buy baby chickens. Chick feed has more nutrients that help chickens grow into healthy egg-layers. Once a chicken starts to lay eggs or reaches 16 weeks, it is time to switch to adult food. Adult chicken food contains oyster shells for calcium, plus other helpful nutrients for a happy, healthy bird. Calcium is imperative for a strong shell. If you do not have a food high in calcium, then the shells will be a weak and possibly lead to too many lost eggs.

If you buy food, Layena by Purina is a good, clean brand. It is affordable with the added plus it is high in oyster shell calcium. Chickens can also be fed table scraps. Free-range chickens will eat insects, pull things from your garden, and get nutrients from the ground. You can supply eggs, rice, vegetables, and fruits for your chickens.

However, realize your chickens are birds of routine. If you feed a store-bought food like Layena crumbles, it may be impossible to change their diet. In one case where the chickens ran out of food and had only pellet food, it was discovered that the chickens wouldn't eat the pellets because it was too much work. These same spoiled chickens also avoided any homemade food.

If You Want Non-GMO, Totally Organic Food, Consider the Following:

1. Alfalfa meal for protein

2. Corn

3. Peas for protein

4. Wheat

5. Oats or Barley (less than 15 percent of the other ingredients)

6. Aragonite, limestone, oyster shell

7. Grit

8. Salt

9. Crab meal

10. Flaxseed

11. Kelp

12. Fish meal

13. Cultured yeast

A mixture of the above ingredients ensures a healthy diet with no commercial made food. Note, the amounts are not provided because you need to make batches based on how many chickens you have and whether you have storage for a few days of food or need to make a mix each day.

Raising chickens for eggs differs from raising them for meat. A meat chicken grows rapidly and eat a lot of protein to grow meaty for you to enjoy. Although you can allow your hens to produce for a couple of years and then slaughter them for a meal, generally the younger the chicken is the better the meat will be. If you buy chicken breeds intended for meat, they are supposed to be slaughtered within a year, or their heart may give out. These are the chickens genetically breed to rapidly increase in size.

Chickens may be the best livestock to start your farm with, but there are six animals you can raise on just a quarter of an acre. If you have a full acre, you can add in more livestock.

5 More Livestock to Keep

The livestock in this category is like the ease of care and space to chickens. If you want small animals for meat or eggs, then the following will be beneficial to your homestead:

1. **Ducks:** Much like chickens, you do not need a huge space to raise ducks. Male ducks do not crow, so they are also allowed in neighborhoods that might be against roosters. Ducks need 4 square feet of coop space per bird and plenty of running around space. Like chickens, you can raise ducks for both eggs and food.

2. **Dwarf Goats:** Goats are multipurpose animals because they can offer meat and milk. One goat requires 5 to 6 sq ft of space, plus 20 feet of grazing area. If you do not want to tackle raising cows for milk or beef, then a goat may be the next best option.

3. **Sheep:** Another multipurpose animal that provides meat, milk, and fiber is sheep. Shearing sheep for wool to spin and make clothes from is an added benefit to raising sheep on your farm. Unfortunately, sheep are more difficult to milk due to their instinctual behavior against predators. A grown sheep needs 12 to 16 sq ft of living area plus 16 to 25 sq ft of grazing area.

4. **Quail:** Quail is a small fowl that requires less space than ducks. Unfortunately, they can be on the noisy side, which is not good for city living. Off-grid dwelling makes a perfect place to raise quail for their eggs and meat. They eat less and require

less bedding, so you can raise more quail per sq ft than chicken or ducks, for less money. However, you get less meat.

5. **Rabbits:** I cannot think of eating Thumper, but rabbits have been food for centuries because they are easy to raise. Rabbits can be destructive creatures. Although many consider it a myth, rabbits can crawl into cars and other tight spaces and chew wires. Rabbits are good for fertilizer, meat, and angora wool. They eat little, and you can feed them natural grasses to keep costs down. You will need to provide at least 2 sq ft per rabbit and up to 5 sq ft for a 12-pound rabbit.

Cattle

Cows or cattle need to be in their own section because a lot goes into raising milk cows and beef cows. First, a beef cow takes at least a year to raise, where you help it grow from a 500-pound calf to a 1,000 or more-pound cow. It takes at least 2 acres to feed a calf and raise it to a cow. One beef cow can feed a five-person family for one year, as long as you have other options like chicken, pork, and fruits and vegetables.

Cows do graze on grass, but you also need to feed additional nutrients. Many cattle owners will provide corn to help their cows gain weight. You also need to provide bales of hay. The formula for figuring out how much hay is to consider a cow needs 3 pounds of hay per 100 pounds of weight. So, a calf that weighs 250 pounds needs about 7 pounds of hay per day. If you have a lot of grassland, your cows may get enough grass by grazing.

The vet fees for taking care of your cattle can also be extensive. Depending on where you live and how close the vet is, you can expect to pay at least $45 for the visit and an additional $40 to $50 for any vaccinations that may be needed. In California, the costs can be higher at a whopping $90 per non-emergency visit. Note: these costs are for a regular visit to ensure your cattle are healthy. Fees for birthing help, illness, or emergency can increase the cost to well over $150.

Angus and Hereford cattle are the best for meat, but there are other choices you can buy.

Pigs

Pigs are another way to raise meat on your farm. A 200-pound pig can provide enough pork, ham, and bacon for a family of four, while two pigs are best for up to eight people. You can have a farm full of pigs that produce baby pigs you can raise for slaughter, or you can continue to buy piglets you raise until it reaches 225 to 325 pounds.

Buying a pig in March or April helps you get the pig ready within a year for slaughter.

You will need to pay vet fees for the inoculation of various diseases. A pig should be inoculated for cholera at eight weeks. You also want to ensure you buy a pig that has already been through a deworming regime. Pigs can be expensive, not only for the grain you buy and the vet visits, but for their full care.

Your pig will need a pen and a house. A simple structure 8 x 6 x 5 feet will suffice.

At 200 days, your pig should be 325 pounds.

There is no specific breed of pig that is better than another when it comes to meat.

Slaughtering Livestock

Being self-sufficient is one thing but having the time and knowledge to slaughter your livestock is completely different. It takes years and more than a short section in a homesteading book to discuss the best methods of slaughter.

1. Make sure the laws of your state allow for the slaughter of livestock. If you intend to do it yourself, the laws may restrict you to where you must find a creditable slaughterhouse.

2. Research slaughterhouses and prices.

3. Assess their methods and the cleanliness of the location.

4. Make sure the process is humane.

5. Seek a professional who can cut the meat correctly and use all parts of the animal.

6. Most places will help provide bacon, which can be hard to make on your own if you slaughter your own animals as a novice.

Protecting Your Livestock

Beyond fencing the grazing land, you want to have a guard dog. Several breeds of dogs are bred to be herders and protectors of livestock. A dog can deter a larger predator by being around and on watch. Any livestock dog will make an excellent pet too, but you want to encourage them to bark only at danger. Not all dogs will make the best guards due to personality differences. Some breeds meant for guarding can also produce a lazy dog, while others in the same litter can be the best guards you will ever see.

It is best to seek a breeder for a non-mixed breed and to be clear in what you want from your puppy. Starting with a puppy ensures that you raise him or her for what you wish to see in behavior.

Chapter 11: Preserving Your Food

Did you grow too many tomatoes? Or maybe it was too many blueberries? What will you do since you don't want to waste the food, but you and your family cannot eat it fast enough? Preserving food solves keeping your items fresh and edible throughout the year or years you take to eat everything. As a homesteader, who wishes to be self-sufficient, learning how to can, ferment, pickle, dry, and freeze various foods guarantees self-sufficiency. Other food preservation techniques discussed are cold storage, using a root cellar, and salt curing and smoking to preserve meats.

Drying

Drying requires a food dehydrator, sun, or shade. Shade or adiabatic drying is a process without heat. Solar or sun drying helps the fruit or vegetables dry in the sun, where the container captures the heat. Apricots, grapes, and tomatoes are the most common foods sun-dried.

A food dehydrator uses air and heat. You can also achieve the same concept by using a warm oven, set at 165 degrees F. Using a food dehydrator is safer because it has a timed setting for the foods you might wish to dry, including jerky. Making jerky in a warm oven can go wrong, causing microorganisms to grow. You want the moisture to evaporate quickly, so there is no possibility of bacteria surviving the process.

Once the food is properly dehydrated, it can be stored up to 12 months, depending on the meat, fruits, or vegetables.

Canning

Canning helps you seal the food to prevent it from turning. Jelly made from fruit is one way of canning. You may also can tomatoes or turn them into spaghetti sauce before sealing the container against oxidation.

The process of canning makes food self-stable if the food is in an airtight, vacuum-sealed condition. It is important for all foods canned to reach a temperature of 250 degrees F during the process to render enzymes inactive and microorganisms dead. As the food cools, in the vacuum-sealed container, it cannot grow new bacteria.

When canning, it is imperative you do so using appropriate guidelines and not a random recipe or method found online. Home-canned foods contain a higher risk of botulism.

Canning your food requires you to leave a headspace, so the proper vacuum forms. Typically, the liquid should stop at the neck of the canning jar, so there is an inch of headspace between the liquid and the lid.

Use any canned foods you prepare within a year. If you feel the seal on any can have been compromised, throw the contents away.

Boiling Water Canners

Boiling water canners are usually porcelain or aluminum. They have perforated racks with fitted lids that help you boil the water to the correct temperature. Visit https://nchfp.uga.edu/ for proper canning instructions. It is a USDA backed PDF that tells you how to go through the canning process for the utmost safety.

You can also buy a canning machine that takes all the guesswork out of preserving your food. You simply put in the food, and the machine does the rest of the work.

Root Cellar

A root cellar helps you store carrots, onions, potatoes, and other root foods in a dark, underground place. Apples and tomatoes can also be put in a root cellar.

Freezing

Freezing prevents something from spoiling until you can use it. You can freeze almost any fruit, vegetable, or meat. You want to keep the food completely frozen and avoid anything that may have thawed and then refrozen. Freezing will not kill microbes like bacteria or parasites, which is why you need to ensure your livestock is vaccinated and that anything you grow is parasite-free. Freezing will not destroy any of the nutrients in the food; however, left too long in the freezer, things can become freezer burned. Although the meat is safe, it's tougher meat so most people cut off the dried bits. The types of packaging you choose determines how long something can remain in a freezer without deteriorating. As soon as you slaughter your livestock and package it, you will want to freeze it to ensure it retains the proper quality. A freezer should be at 0 degrees Fahrenheit for the proper temperature of the items inside.

When you thaw the food, do so in the refrigerator or in chilly water. It is better to thaw it in the fridge overnight rather than leaving it on the counter or in water that will warm. It is also possible to use a crockpot and thaw the meat as it is cooking for 8 hours.

Meat, uncooked, will last in a freezer for four to twelve months. Bacon, sausage, ham, hotdogs, and lunchmeat are good for up to two months. Casseroles, gravy, poultry, cooked meat, soups, and stews are good up to four months. Egg whites, uncooked chicken, and wild game can last for a year.

Pickling

Pickling entails putting something in a vinegar liquid to keep them longer. Pickles made with cucumbers through the pickling process are the most common. However, cabbage, radishes, jalapenos, banana peppers, and other vegetables can be pickled to preserve them for use later.

Salting

Salting is a method of preservation used for centuries. One of the most typical meats preserved through salting is ham. When ham is salted it can last for more months than if it was left uncooked.

You do not have to cook the ham for it to become dry-cured and remain stable for several months on the shelf. However, you do need to make sure you have used enough of the salt and other ingredients to successfully remove all moisture from the ham. Dry-cured hams take a few weeks to a year to age properly. Six months is the typical amount of time a dry-cured ham requires for the correct taste and softness so it's ready to eat.

Smoking

Smoking, which uses a wood smoker, helps you cook the meat with flavor and ensure it lasts longer than if it was frozen uncooked. Smoking can be done in a wood smoker or over a fire that slowly smokes the meat. You want wood chips that will provide a flavor to the meat. Check for the correct temperature of smoked meat since ham, chicken, and beef will need different temperatures before they are considered done. Smoking is like salting because you are dehydrating the meat to prevent bacteria from growing. The reason you might wish to use smoking over salting is the salt. Too much salt is never good for the diet, so using hickory, oak, cherry, maple, or applewood chips ensures you have a nice flavor without too much salt. Smoking generally requires eight hours, if not a full 24 hours. The longer you smoke the meat, such as 48 hours, the longer you can keep the meat.

Chapter 12: Making Money From Your Homestead

Once your homestead is up and running, it is time to use your resourcefulness in gardening, farming, animal husbandry, beekeeping, and off-grid living to turn a profit and supplement your needs. There are a variety of ways you can earn money by selling your oversupply of vegetables and fruit, and selling other items you raise or make. Farmer's markets are a wonderful place to offload extra food, and town festivals give you a chance to sell things like homemade soap.

There are at least 50 ways you can make money from your homestead.

Money from Animals and Insects

The following will discuss how you can make money from animals and insects you raise on your homestead.

1. Bees

Becoming a beekeeper helps you make money because of the honey you can collect from honeybees. In a time when honeybee populations are endangered, it is more important for homesteaders to do their part in raising colonies and enjoying the benefits that come from bees.

- Start small, such as one honeycomb.

- Seek a professional's advice on how to approach the bees and collect honey so you may avoid endangering your life or the bees.

- Collect the honey for yourself and sell the extra.

2. Breed Livestock

Homesteaders have not only raised cattle and other livestock for meals, but also for sale. You can breed livestock to sell to other homesteaders just getting a start, or you can sell the beef you gain from raising livestock.

There are three ways you can make money from livestock.

- Breed them for sale.

- Breed them for slaughter and sell the meat.

- Become a livestock consultant. By enticing people to your homestead to learn from you and gain advice, you can make money to help feed your livestock.

- Breed Livestock Guard Dogs.

If you have a great pair of guard dogs that help protect your animals, there is no reason you can't breed them and sell their offspring. Finding the correct personality for watching over livestock is not always easy. Breeders create dogs for many reasons, including pets. Finding a breeder that provides dogs for guarding livestock are fewer.

Most purebred dog breeders can get a couple of hundred dollars or more for their dogs.

3. Making Products to Sell

Making your own products and selling them is also time-honored in the homesteading life. Soap is one of the biggest ways to make money online and at art fairs. Castile, goat milk, olive, coconut, shea butter, and lye-based soaps are something our ancestors had to make. Typically, the original soap was made with lye, but now you can use things like goats' milk by raising goats.

One of the things that are lacking in the world is homemade soap without additive fragrance or perfume. Using natural oils, such as mint from your mint plants, is much better than any fragrance you could buy and use. A huge movement is underway right now to use natural, organic soaps that lack chemicals but use natural oils to ensure healthy skin.

Soap is just the beginning. You can move on to shampoo, conditioner, lotion, lip balm, and other toiletries.

With vegetables like avocado and herbs such as mint, you can add ingredients designed to help the hair and skin become healthier. Several herbs and plants you grow can have natural remedies for dry skin, aging, and even pain.

Consider making products with antioxidants, anti-inflammatory, and anti-bacterial properties as these will also sell due to the natural ingredients.

4. Firewood

Cutting down healthy trees is obviously not something you want to do, and depending on where you put your homestead, you may not have a lot of trees. However, for those who live near forests like Colorado, fire mitigation is essential to keeping a homestead safe, and it provides plenty of firewood for you and potentially others.

Anytime you clear land of trees or conduct fire mitigation, there is nothing wrong with selling bundles of wood to those who needed.

If you do not want to cut the firewood, you can also offer a lesser cost for a person to come and chop firewood for their personal use. If you already have downed trees, then paying someone to haul away the trees for firewood can get you a little extra funding.

5. Growing and Selling Vegetables/Fruit

Anytime you have an overabundance of veggies and fruit, even after canning some to save for the year, you can get a booth at a farmer's market and sell your goods. You want to choose the best items to bring and have comparable costs to those selling around you.

Another option is to get a license that allows you to sell to a restaurant. Restaurants in certain areas are always happy to pay a reasonable price for homegrown mushrooms.

6. Freelance Work

As a homesteader, there are a lot of things you have learned or perhaps already knew. You can use these technical skills to make money. Writers use sites like Upwork to connect with clients looking for books about homesteading. If you don't want to share your knowledge for a few cents per word, you can always author a book and self-publish on Amazon and other online sources, like Bookshop.org. You will need marketing skills, but you can make money from your knowledge.

Offering your farming skills, woodcutting, or property for livestock are other ways you can freelance for profit.

7. Making Clothing

Remember those rabbits with Angora wool and the sheep and goats with their fur? For anyone with the skills, sewing and knitting clothing, towels, and blankets are ways to make money. People love homemade items of quality and will be willing to pay for them. Like the soaps, you can make your clothing and other items to sell at art festivals or create an online website where you sell whatever you make.

8. Candles

Candles are other homemade commodities that people will purchase. Plus, making candles is actually easy. You need wax, wicks, and a container. You will also need a pot to melt the wax in and color or oil to make the candles beautiful. A mason jar is a good container because it is clear. It also provides a base for the person to continue burning the candle until finally it is gone.

As with soaps, finding non-perfume or fragrance laden candles is hard. Essential oils are a better option or no scent at all. In fact, there is an entire untapped population who would buy candles if they were just pretty in color and lacked any oil or scent. This population has an allergy to fragrance-based smoke, and there are more of us out there than you might imagine.

9. Bread

Making bread from the yeast and other ingredients is possible. French, peasant, and Amish bread are just a few options. Here is one recipe you might want to try:

Amish Bread

2/3 cup sugar

2 cups water (warm)

1 ½ TBLs yeast

¼ cup oil

5 to 6 cups flour (all-purpose)

As a rising bread recipe, you will need at least four hours for the bread to double in size. Then, you will need to knead it, and let it rise a second time before baking it at 350 degrees F.

Any bread recipe you find will add to your table, but also to your pocketbook because you can make several loaves of bread to sell at farmer's markets alongside the fruits and vegetables you grow.

10. Cheese

Goat cheese is another way to earn money, but there are also plenty of other types of cheeses you can make and sell at markets.

11. Selling Eggs

Selling your chicken quail and duck eggs are another three ways you can make money from your farm. By selling the eggs, you also raise money for the feed it takes to keep your flock healthy and productive.

People will pay at least $4.00 per dozen for fresh, free-range eggs.

12. Starting an Orchard

Depending on where you live, it may be possible to grow grapes, apples, and other fruit trees. You not only benefit from the fruits grown, but you can also open your orchard to those who may wish to come pick their own fruits. As a seasonal pick location, it is possible to make more money.

13. Raising Bait

Do you live in an area with a lot of fishermen? Do they need bait? There is no reason you cannot start a worm or bait farm that would make you money. Worms are just one type of bait you might raise. You can also raise fish and use their eggs for bait to sell to local fishermen.

14. Renting Your Property

Do you have some land you could build a little cabin on? Perhaps you want to take a little break from the homestead? By putting your home on Airbnb you can rent it out and make a little money. If you have an area for camping or RV storage, you can also make a little money by collecting a small renter's fee.

There are at least 50 ways to make money from your property. While only 15 have been mentioned, just think of what you can do with your skills. Perhaps you made your own furniture, or do you know how to make cabinets? Maybe you are great with pickles, jams, jellies, and syrups. Any skill you must feed your family and keep them sheltered can turn into more than a hobby and a way of life—it can become a method to make money from your land and skills.

Conclusion

Congratulations on reaching the end of this guidebook. You have it to refer to as many times as you need to become successful in your off grid living choice. Whenever you have doubts or questions, let this book be your answer to moving from novice to expert homesteader.

You knew the general definition of off grid living, and now you understand that you can choose how extreme you are and whether you raise livestock or take advantage of a nearby farm.

You learned you don't have to be off grid without proper water, electricity, and sewage disposal.

It is up to you to take advantage of the different chapters and set up a homestead that will work for you.

Remember, your commitment to being off grid, even with the Internet, will determine your success. Start small and make your way towards larger goals such as raising chickens before getting a cow to milk or raise for beef.

Your first step, now that this book is complete, is to either find the property for your homestead or to research where you can live comfortably. Do not forget your due diligence in researching the city, county, and state laws so you may have a successful homestead and perhaps a business one day.

Living off the land like our ancestors is not only rewarding, but an essential means of survival.

Here's another book by Dion Rosser
that you might like

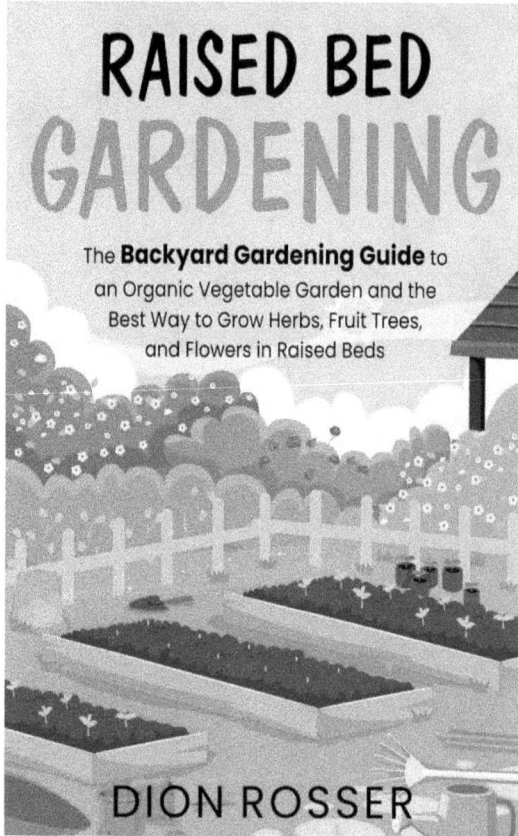

RAISED BED
GARDENING

The **Backyard Gardening Guide** to
an Organic Vegetable Garden and the
Best Way to Grow Herbs, Fruit Trees,
and Flowers in Raised Beds

DION ROSSER

References

10 off-grid homes for a self-sufficient lifestyle. (2019, February 16). Dezeen. https://www.dezeen.com/2019/02/16/off-grid-self-sufficient-homes-sustainable-architecture/

103 Ways to Make Money from Your Homestead, even a small one. (2019, April 16). The Rustic Elk. https://www.therusticelk.com/103-ways-make-money-from-your-homestead/

Aaron, P. (2018, December 20). *Off Grid House Plans.* The Simple Prepper. https://www.thesimpleprepper.com/prepper/homesteading/off-grid-house-plans/

admin. (n.d.). *Off Grid Living.* Http://Off-Grid-Living.com. Retrieved from https://off-grid-living.com/getting-started-off-grid/

All About Off Grid Wastewater: Options, Septic, Code, and Advice. (2017, July 25). Accidental Hippies. https://accidentalhippies.com/2017/07/25/off-grid-waste-septic/

Beckett, T. (2019, May 30). *What is Off Grid? | Living Off The Grid In Rural or Urban Areas Explained.* Off-Grid Expo. https://offgridexpo.com/what-is-off-grid/

Boni, T. (2008, October 10). *Living Off the Grid: How to Generate Your Own Electricity | Today's Homeowner.* Today's Homeowner. https://todayshomeowner.com/living-off-the-grid-generating-your-own-electricity/

Contributor, S. L. (2014, March 12). *Off The Grid Gardening Tips.* Survival Life. https://survivallife.com/off-grid-gardening-tips/

Eric. (2014, December 17). *Living Off The Grid: What Does It Mean?* Off Grid World. https://offgridworld.com/living-off-the-grid-what-does-it-mean/

eric. (2019, June 9). *The Realities of Living Off Grid.* Bubba On The Road. http://www.bubbaontheroad.com/2019/06/09/the-realities-of-living-off-grid/

finally time to think: pros and cons of living off the grid - cope with life. (n.d.). Retrieved from https://www.copewithlife.ca/introvert/finally-time-to-think-pros-and-cons-of-living-off-the-grid/

Food Preservation. (n.d.). New Life On A Homestead | Homesteading Blog. Retrieved from https://www.newlifeonahomestead.com/the-homestead-kitchen/food-preservation/

Food Storage and Preservation | Nutrition.gov. (n.d.). Www.Nutrition.Gov. https://www.nutrition.gov/topics/food-safety/safe-food-storage

Gardening Off the Grid: Mastering Water and Solar Systems. (2018, May 29). Hobby Farms. https://www.hobbyfarms.com/gardening-off-the-grid-irrigation-water-solar/

Getting Started With Off-Grid Water Systems For A More Self-Reliant Homestead • Insteading. (n.d.). Insteading. https://insteading.com/blog/off-grid-water-system/

H, S. (2015, September 7). *Underground Houses: The Ultimate In Off-Grid Living?* Off The Grid News. https://www.offthegridnews.com/extreme-survival/underground-houses-the-ultimate-in-off-grid-living/

Homestead Survival Site - How to Live Off The Grid in Comfort and Style. (n.d.). Homestead Survival Site. Retrieved from https://homesteadsurvivalsite.com/

How to Design an Off-Grid House - GreenBuildingAdvisor. (2017, June 2). GreenBuildingAdvisor; GreenBuildingAdvisor. https://www.greenbuildingadvisor.com/article/how-to-design-an-off-grid-house

How To Get Started Living Off Grid The Homesteading Hippy. (2014, November 3). The Homesteading Hippy. https://thehomesteadinghippy.com/off-grid-living/

How to Preserve Meat in the Wild Without Refrigeration. (n.d.). Know Prepare Survive. Retrieved from https://knowpreparesurvive.com/survival/how-to-preserve-meat-in-the-wild/

Hunter, J. (2016, October 19). *Off Grid Checklist: Become Self-Reliant with these Steps.* Primal Survivor. https://www.primalsurvivor.net/off-grid-checklist/

Instructables. (2014, June 30). *DIY OFF GRID SOLAR SYSTEM.* Instructables; Instructables. https://www.instructables.com/id/DIY-OFF-GRID-SOLAR-SYSTEM/

Living Off the Grid: Pros and Cons - GeekExtreme. (n.d.). Retrieved from https://www.geekextreme.com/design/living-off-grid-pros-cons-20519/

Living off the Grid: Starting From Scratch (Part 1). (2014, November 1). The Good Men Project. https://goodmenproject.com/featured-content/kt-living-off-the-grid-starting-from-scratch/

Living Off-Grid: What It's Actually Like • Insteading. (n.d.). Insteading. Retrieved from https://insteading.com/blog/living-off-the-grid/

Max, A. be ready". (2019, October 14). *How To Go Off-Grid Step-by-Step.* American Patriot Survivalist. https://americanpatriotsurvivalist.com/how-to-go-off-grid/

MorningChores - Build Your Self-Sufficient Life. (n.d.). MorningChores. Retrieved from https://morningchores.com

MOTHER EARTH NEWS | The Original Guide to Living Wisely. (n.d.). Mother Earth News. Retrieved from https://www.motherearthnews.com

Oetken, N. (2018, December). *10 Ways to Preserve Meat Without a Fridge or Freezer.* Urban Survival Site. https://urbansurvivalsite.com/ways-preserve-meat/

Off Grid World - How To Live Off The Grid. (n.d.). Offgridworld.com. Retrieved from https://offgridworld.com/

Off-Grid or Stand-Alone Renewable Energy Systems. (2020). Energy.Gov. https://www.energy.gov/energysaver/grid-or-stand-alone-renewable-energy-systems

Pacific Lutheran University BA, E., & Twitter, T. (n.d.). *Generating Off-Grid Power: The 4 Best Ways.* Treehugger. Retrieved from https://www.treehugger.com/generating-off-grid-power-the-four-best-ways-4858714

Pedersen, D. (2018, May 29). *Realities of off-grid living.* The Land. https://www.theland.com.au/story/5435370/realities-of-off-grid-living/

Primal Survivor. (2017, February 2). Primal Survivor. https://www.primalsurvivor.net/

Pros and Cons of Living Off Grid. (2019, September 1). Offgridmaker.com. https://offgridmaker.com/2019/09/01/pros-and-cons-of-living-off-grid/

R. Paul Singh, & H. Russell Cross. (2017). Meat processing - Livestock slaughter procedures. In *Encyclopædia Britannica.* https://www.britannica.com/technology/meat-processing/Livestock-slaughter-procedures

Raising Chickens Off the Grid ~ Without Heat Lamps or Lights. (2018, April 30). Practical Self Reliance. https://practicalselfreliance.com/chickens-without-electricity/

Schwartz, D. M. (n.d.). *Best Alternative Off Grid Toilets - No Septic!* Off Grid Permaculture. Retrieved from https://offgridpermaculture.com/Water_Systems/Best_Alternative_Off_Grid_Toilets___No_Septic_.html

The Reality of Living Off-Grid in a Caravan with Children in Central Portugal Over Winter. (2018, February 25). Topsy Turvy Tribe. https://topsyturvytribe.com/portugal/the-reality-of-living-off-grid-in-a-caravan-with-children-in-central-portugal-over-winter/

What Is an Earthship Home? Eco-Friendly Living and Zero Utility Bills. (2018, January 24). Real Estate News and Advice | Realtor.com®. https://www.realtor.com/advice/buy/earthship-home/

What is The Meaning of Living Off The Grid? (2020, January 15). An Off Grid Life. https://www.anoffgridlife.com/what-is-the-meaning-of-living-off-the-grid/

www.ingramcontent.com/pod-product-compliance
Lightning Source LLC
Chambersburg PA
CBHW050644190326
41458CB00008B/2420